CLIMATE TALK:
RIGHTS, POVERTY AND JUSTICE

CLIMATE TALK: RIGHTS, POVERTY AND JUSTICE

Edited by
Jackie Dugard, Asunción Lera St. Clair and Siri Gloppen

JUTΛ

Published 2013

© JUTA & Co Ltd
First Floor,
Sunclare Building,
21 Dreyer Street,
Claremont 7708
lawproduction@juta.co.za

ISBN 978 1 4851 0064 5

TYPESETTING: ANDTP SERVICES, CAPE TOWN
PRINTED AND BOUND BY

TABLE OF CONTENTS

CLIMATE TALK: RIGHTS, POVERTY AND JUSTICE – INTRODUCTION

Climate change is one of the central challenges facing African countries and their people. Unless concerted efforts are made worldwide very soon to reduce emissions, climate change impacts in Africa are likely to be devastating.

The Fourth Assessment Report of the Intergovernmental Panel on Climate Change (IPCC) documented that development and poverty reduction achievements are at serious risk if average temperatures rise above the two-degree mark.[1] Over the past five years, scientific scholarship has produced even harsher prognoses. The possibility of higher-end temperature scenarios suggests a dark future in which humans will struggle to ensure a healthy environment and to access basic needs such as access to water, food and secure housing.[2] Changes in the onset of seasons, spread of disease vectors (such as malaria and dengue), reduced energy access and decreased water availability, as well as increased frequency of extreme weather events and associated damage to infrastructure and housing, will deepen existing human vulnerabilities.

Such changes also add to the stressors on our natural resources, combining with other human pressures such as unsustainable land use, deforestation and poor management of water resources.[3] All of these problems are present in Africa and are likely to be exacerbated by climate change.

As established by the IPCC Special Report on Extreme Events, the capacity to respond to extreme or even small weather events such as storms, floods, or

---

1   IPCC 'Fourth Assessment Report: Impacts, Adaptation and Vulnerability' (2007) <http://www.ipcc-wg2.gov/publications/AR4/index.html>.
2   See M New, D Liverman, H Schroeder & K Anderson 'Four Degrees and Beyond: the Potential for a Global Temperature Increase of Four Degrees and its Implications' (2011) 369 *Philosophical Transactions of the Royal Society of London* 6; S Smith, ML Horrocks, A Harvey & C Hamilton 'Rethinking Adaptation for a 4oC World' (2011) 369 *Philosophical Transactions of the Royal Society of London* 196; World Bank 'Turn Down the Heat: Why a 4 Degree Warmer World should be Avoided' (2012) <http://climatechange.worldbank.org/sites/default/files/Turn_Down_the_heat_Why_a_4_degree_centigrade_warmer_world_must_be_avoided.pdf>.
3   See PH Gleick *Bottled and Sold: the Story Behind our Obsession with Bottled Water* (2010); IPCC 'Special Report on Managing the Risks of Extreme Events and Disasters to Advance Climate Change Adaptation' (SREX) (2012) <http://ipcc-wg2.gov/SREX/>; J Rockström, W Steffen, K Noone, Å Persson, FS Chapin III, E Lambin, TM Lenton, M Scheffer, C Folke, H Schellnhuber, B Nykvist, CA De Wit, T Hughes, S van der Leeuw, H Rodhe, S Sörlin, PK Snyder, R Costanza, U Svedin, M Falkenmark, L Karlberg, RW Corell, VJ Fabry, J Hansen, B Walker, D Liverman, K Richardson, P Crutzen & J Foley 'Planetary Boundaries: Exploring the Safe Operating Space for Humanity' (2009) 14(2) *Ecology and Society* 32 <http://www.ecologyandsociety.org/vol14/iss2/art32>; World Health Organization (WHO) 'Climate Change and Health' (2010) <http://www.who.int/mediacentre/factsheets/fs266/en/>.

droughts is determined not only by the magnitude of the natural event, but by the socio-economic and political conditions of the regions and communities affected.[4] For example, overcrowded cities with high levels of poverty and poor housing and sanitation conditions are much more vulnerable to any weather.[5] The report highlights the fact that in the past years 90 per cent of all deaths due to extreme weather events occurred in developing countries.

Thus it is clear that the groups that will suffer the most from the challenges posed by climate change are those that are already suffering the negative impacts of other global challenges, such as the financial/economic crises, ongoing conflicts, environmental degradation and loss of biodiversity; those without voice or power, assets or access to energy, lacking insurance for flooding or for destruction caused by severe weather events; those with poor or no access to health, education, clean water and under conditions of food and labour insecurity; and those whose agricultural systems have already been negatively affected by global trade and patent regulations.

To make matters worse, these are the same groups that have contributed the least to the climate change problem in the first place: the poor and vulnerable people around the world, and especially in less-developed countries, who have benefited the least from modernisation and industrialisation and have a relatively small carbon footprint. This is a double injustice.

From the perspective of justice, therefore, climate change demands bottom-up analyses of climate impacts, along with ones that link the narratives and scientific information presented in global and domestic documents with debates in relation to existing rights systems and poverty strategies. Yet, there has been relatively scant work focused on framing climate change in relation to the multiple challenges it poses for poor people and their rights. And, notwithstanding a reality in which climate justice and social justice are very difficult to disentangle, neither the legal systems nor the main actors framing the dominant climate change narratives seem to be sufficiently attentive to the double-edged justice questions raised by climate change impacts on poor communities in Africa (and beyond).

Such deficits in the framing of climate change have stunted concerted action. So, even though the science is clear and there is broad agreement of the magnitude of the threats and need to act urgently, there is little action of practical import towards addressing the violations and vulnerabilities of the poor. Indeed, despite being held in Africa (the poorest and most vulnerable continent), the United Nations Framework Convention on Climate Change (UNFCC) Conference of Parties 17 (COP17) held in Durban in 2011 was an all-too-familiar display of the impotency of global leaders to respond adequately to the looming crisis.

Why is it so difficult to translate knowledge into action in the field? Many factors contribute. Climate change is a difficult problem with complex interdependencies and radical uncertainty about effects of different measures. The need for long time-horizons and solutions that transcend national

4   IPCC ibid.
5   Ibid.

boundaries renders existing decision-making structures poorly suited to deal with the problems. The potential for producing sustainable and equitable solutions is further diminished by unequal power relations and divergence of economic and political interests. But the paralysis is also related to how we frame and understand the issues, which narratives and discourses emerge and become dominant, and whose voices are heard and included in decision-making processes.

In this book, we attempt to fill some of the gaps in climate change scholarship by focusing on the climate narratives and discourses emerging in and around South Africa – how they relate to broader issues of social justice and resource allocation, and the role of rights talk and legal strategies in the framing of the problems and solutions.

The authors – all scholars and/or activists who are currently involved in building climate justice – grapple with a range of discourses and practices, including international development discourse, legal and rights-based strategies, and emerging green growth policies. They ask a range of questions such as: what narratives on poverty are being 'used' by different actors in relation to climate change? How are national level narratives influenced by global discourses – and how do they differ? Which claims have more epistemic and political weight and why? Whose voices are privileged and whose voices are silenced? How do activists on social rights and poor people's empowerment incorporate and relate to emerging concerns with climate change? How do climate change narratives of various voices in the development debate relate to scientific knowledge on climate change? How is climate change science knowledge used by political actors? What are the central social justice and ethical dimensions of climate change? How can lawyers and courts mediate environmental and climate change issues? And how are these questions dealt with by key global actors in relation to South Africa and by South African actors in relation to both domestic and international processes?

The chapters presented here approach these questions from very different perspectives. These include frame and policy analysis, ethnographic and legal analysis, linguistics, media studies, political economic and climate justice perspectives. Most of the chapters have in common the integration between fairness related to environmental issues and fairness related to socio-economic issues, filtering these justice connections through the additional layer of climate change.

The first section of the book 'Talk is cheap' addresses the relations between climate change, justice, rights and the poor, and the narratives that construct these relations, in the context of formal policy and ethical frameworks. The first chapter, by Kjersti Fløttum and Øyvind Gjerstad, analyses the official South African government discourse on climate change, as is manifested in the National Climate Change Response White Paper of 2011. Combining linguistic and discourse analysis, the authors analyse the White Paper from a rights and poverty-related perspective by examining the occurrence, distribution frequency and linguistic representation of social justice related terms such as socio-economic rights, vulnerability and poverty. The analysis shows that, while 'poverty' appears a lot in the text, in the White Paper the legal rights of the poor are hardly given any place in the argumentation, with the government framing the fight against poverty in terms of intentions rather than duties. And, despite many (polyphonic) voices indicating a complex 'plot' in which the South African government is both the victim and the villain in terms of global emissions (itself an indication of South Africa's highly unequal society), a conclusive engagement with South Africa's industrial policy and its impact on climate change, especially from the perspective of the poor, is absent from the White Paper.

The examination of South African government narratives is complemented from a global perspective in an analysis of the United Nations Development Program (UNDP) Human Development Report (HDR) 2011 on sustainability and equity, by Des Gasper, Ana Victoria Portocarrero and Asunción Lera St. Clair. Here, the authors investigate how the UNDP's thinking on environment and sustainability has evolved since the HDR 2007/2008 on climate change and development. This follows up on a previous article by the authors, in which they show how the 2007/2008 report made substantive arguments regarding the importance of considering human rights and the interests of the poor when framing the relations between climate and development but refrained from bold policy proposals, falling back on recommendations promoting market and technocratic solutions.[6] The current chapter highlights that the HDR 2011 moves further in a technocratic direction. Largely apolitical and insensitive to human rights and justice issues, the HDR 2011 constitutes a diluted successor to the HDR 2007/2008, and closer in perspective to the World Bank. It is focused on technocratic win-win-win synergies between the environment, equity and human development, where the latter is viewed in narrow, instrumental terms. It downgrades attention to future generations, the poor, human rights, and the responsibilities of rich countries. Moreover, it offers a re-framed and weakened version of sustainability compared with the traditional framing of the Brundtland Commission, leading to less commitment. The 2011 report's weak treatment of the absolute importance of climate change and the socio-economic and political barriers to sustainability and to climate change impacts, leads to an insufficient and sometimes misleading picture of the problems faced by the most vulnerable sectors of South Africa's society. This climate change narrative will have serious consequences for sub-Saharan Africa, including for South Africa and especially the poorest social groups.

The final chapter in this section, by Petra Tschakert and Nancy Tuana, tackles the ethical considerations of the climate change debate. As the authors point out, within climate change debates, three concepts – vulnerability, resilience and human security – have been used and abused, resulting in confusion and frustration within the academic community and among practitioners. Re-evaluating these three concepts, the chapter proposes a new interpretation that is focused on the situatedness, relational interdependencies and partiality of resilience. Here, the reframing is based on a relational ontology

---

6   D Gasper, AV Portocarrero & AL St. Clair 'The Framing of Climate Change and Development: a Comparative Analysis of the Human Development Report 2007/8 and the World Development Report 2010' (2013) 23 *Global Environmental Change* 28.

that involves the recognition and prioritisation of the interconnections and interdependencies between people across geographic regions, between current and future generations, and between people and nature. Such a relational ontology provides a resource for reframing the vulnerability versus resilience debate by demonstrating that it rests on a fallacious set of dichotomies, and by requiring a more appropriate conception of human security.

The second section of the book 'COP-ing out' provides a critique of the supposedly popular processes around COP17, held in Durban in November 2011. In the first chapter, Jill Johannessen uses discourse and media analysis to examine the way that climate change was framed in the South African media before and during COP17. Pointing to the moral and justice-related issues entailed by climate change having been caused by rich industrialised nations, while the negative effects are mostly experienced by poor people in developing and poor countries, the chapter investigates from a climate justice perspective how the South African media constructed representations of climate change during COP17. As noted, the mass media is a central arena for representing sometimes-conflicting knowledge claims and narratives and thus constitutes an important site for the production and reproduction of information. Given the scale of the problems and the urgency of the need for solutions, it might be expected that there would be substantial engagement in the media with the underlying issues including poverty and human rights. However, globally, there has generally been relatively scant coverage of climate change issues, particularly from a justice perspective. Examining coverage of COP17 across the South African public broadcaster's evening news on *SABC3*, as well as two daily newspapers (*Business Day* and *The Mercury*), the chapter concludes that the South African media tend towards reflecting scientific information and technocratic solutions over human rights, equity or justice-related framings. Thus, despite extensive coverage of COP17 and notwithstanding some nuances between the media outlets, the South African media's coverage of COP17 overwhelmingly failed to frame climate change as a human rights issue, suggesting the need for further reflection on the role of the media and its engagement with climate change.

The other two chapters in this section continue with the theme of COP17, albeit from the perspective of inofficial spaces and discourses, asking how representative the official spaces and voices were, and providing rare windows into how people on the ground, along with various activist groups present during COP17, framed the challenges of climate change in their narratives, and the solutions presented in these discourses. Brandon Barclay Derman's chapter describes the civil society scene on the outskirts of the official negotiations, outlining how groups operating at transnational, pan-African, national and local scales, congregated in Durban to mobilise for climate justice. Rural women and other stakeholders facing climate impacts were brought in a climate caravan across thousands of kilometres to Durban to represent their interests, transnational activists 'occupied' COP17 to get their message across, and 17 local activist groups organised a People Space. Based on his observations of the range of activities, Derman shows how various groups of activists framed their concerns, finding that demands by civil society for ambitious and binding mitigation in developed countries, and climate finance for Africa were frequently linked with struggles for social justice, accountability, political representation, and survival. He also shows how they highlighted structural inequality and social division at national as well as global levels, and how national groups like the South African Right2Know campaign linked its domestic transparency initiative with opposition to closed-door deal-making in COP17. Derman notes that rights discourse frequently appeared in mobilisation for climate justice during COP17, but concludes that it often played mainly a supporting role within messages, actions, and campaigns highlighting inequality and social divisions at international and national scales.

Molefi Mafereka Ndlovu has a more critical perspective of the climate change activists and non-governmental organisations (NGOs) congregating in Durban, which he sees as 'a tamed civil society that has become a rehearsed spectacle, acting more as an accomplice to power rather than a challenge to the established hierarchies of power'. Based on a series of 'pavement broadcasts' made in Durban during the months immediately prior to and during COP17, which were collated as part of an innovative oral history project titled 'Qwasha!', Ndlovu relates the narratives, songs and re-tellings of ordinary people and grassroots activists (including some of the rural women who were part of the climate caravan described in Derman's chapter). Dialoguing with various members of South African civil society about their understanding of the climate change phenomenon, their views on the formal COP17 process, and the role of the official civil society space, Ndlovu notes a double exclusion – common people and grassroots activists, not only felt excluded from the officialdom characterising COP17, but also from real inclusion in the civil society spaces ostensibly facilitating participation and critical expression for ordinary people. Some respondents vented frustration and a feeling of being used by professional NGOs towards their own agendas, and then discarded 'like used condoms'. Commenting on a relative absence of rights-based climate change mobilisation, Ndlovu observes that most of the grassroots activists interviewed, framed climate change within a general critique of capitalism as a socio-economic system, with its consequent inequalities and injustices. Hence climate change is seen primarily in relation to other pressing mobilisations, such as against privatisation of water or land evictions highlighting the gaps between the rich who benefit and are responsible for climate change, and the poor whose suffering is exacerbated. When rights discourses are engaged, it takes the form of appeals to social and economic rights, rather than the rights of Mother Earth, which figured prominently in relation to COP16, held in Mexico a year earlier.

While anchored in South African realities and experiences, the third section 'Money speaks' looks beyond the COP17 and South African context, towards global processes and discourses. Patrick Bond's sole-authored chapter addresses some of the limitations of rights-based perspectives through a critical analysis of what the author characterises as the neo-liberal takeover of rights language, using the example of a well-known case on water rights (*Mazibuko*). The chapter argues that, in order to serve the principles of climate justice, climate change narratives must be closely monitored in the real world of action and through close observance of the corporate control of natural resources and associated commodification of natural resources. The chapter expands on the classical critique of rights-based perspectives such as the overemphasis on liberal individualism and disconnection from broader socio-economic and ecological processes – and applies these insights to the current debate on climate justice and emergent discourses on the greening of economies with associated narratives on what types of access and management of natural resources is needed at times of a changing climate. According to Bond, precisely because of a history of abuse and appropriation of narratives on rights in relation to nature, it is of central importance to assure a stronger countervailing 'decommodification' narrative than 'rights talk' can offer. The chapter argues for a strengthening of the narrative of commons and the common good, to counter the tendency of appropriation of rights talk.

Khadija Sharife, together with Patrick Bond, analyses the assumptions embedded in the narratives of green growth and payment for ecosystem services. These terms, the authors claim, can be easily used to commodify even more nature and the people whose lives depend directly on nature rather than lead to juridical accountability for the misuse and undervaluing of natural resources. Ecological modernisation is unlikely to lead to climate justice, on the contrary, the methodologies that value nature, primarily cost-benefit analysis, and the unilateral focus on market tools for a transition to sustainability are not able to move the debate beyond the rationality of profits and privatisation of both nature and people. The chapter critically evaluates the failures of schemes such as the Clean Development Mechanism, carbon markets and the failures of the COP and global players to produce agreements on mitigation. But the greatest flaw identified by the authors is the way in which all these market mechanisms remain blind to the large ecological debt owed to poor countries and in particular to African poor communities for historical and present abuses of natural resources without appropriate compensation. The second part of the chapter is devoted to arguing for the primacy of *ecological debt* as a guiding principle for climate justice and elaborates ways in which this idea can become embedded in legal systems. The authors recognise the limits to justiciability in human rights terms and the narrow interpretations of the law in regards to rights claims and offer many examples of the limits achieved in courts so far. The chapter concludes by proposing the idea of a Basic Income Grant (BIG) as the most effective tool to pay for the ecological debt incurred by wealthy countries and social sectors to those less advantaged.

The final section of the book 'Climate justice in court' brings the discussion of climate justice into the realm of the courts. In their chapter, Jackie Dugard and Anna Alcaro focus on environmental justice in South Africa and the relationship between environmental and socio-economic rights, as mobilised by activists and adjudicated by the courts. Under apartheid, the environment was regarded as the preserve of the (white) middle class as conservation of the natural landscape, but never understood as the ecosystems sustaining the lives of poor (black) South Africans. In contemporary South Africa, there is a growing environmental justice movement that seeks to advance a holistic interpretation of the environment as infused with socio-economic rights, and sometimes uses the courts as one of its battlegrounds. Yet, although there have been some important judgments that have begun to look at broader issues, to date, environmental litigation has focused more narrowly on procedural and administrative justice aspects, while leaving the underlying determinants of the problem intact. The chapter suggests that, as climate impacts become more visible, mobilisation for climate justice increases and cases enter the judicial system, activists, lawyers and the courts will have to grapple more meaningfully with environmental justice issues in order to generate a more progressive and all-embracing interpretation of environmental law.

The final chapter, by Catalina Vallejo and Siri Gloppen, re-opens the focus onto the global scene, examining the experience of climate justice-related litigation across the world. As noted by the authors, climate change presents enormous governance challenges, and the generalised failure of political bodies internationally to regulate for effective and fair mitigation and adaptation has sparked alternative strategies from activists trying to force action. This includes lodging cases in courts and quasi-legal bodies at national and international level. Giving an overview of a selection of such cases, the chapter examines how narratives on climate change play out, focusing in particular on whether and how environmental (green) issues combine with socio-economic and cultural (red) issues. As the authors highlight, and echoing the conclusion of the chapter by Dugard and Alcaro, although already marginalised communities are hardest hit by the negative effects of global trade and now climate change, 'red' issues are seldom explicitly addressed in climate lawsuits. Yet, litigating climate change in closer connection to economic, social and cultural rights might allow for structural legal remedies that better come to grips with some of the pressing problems where equity and environmental considerations are interlinked, for example in relation to global energy production.

By presenting this collection of chapters – displaying very different narratives on the challenges climate change poses for the poor and their rights, in South Africa and globally – we have sought to provide a prism with which to view the social and political complexity of the climate change problem. We see in all the chapters critiques of a generalised tendency to shy away from directly addressing the many profound justice-related questions raised by climate impacts on poor and disadvantaged populations. There is clearly a prioritisation of technocratic views, a focus on easy solutions or the low hanging fruit of technical and small adjustments. In many of the official spaces, forums and discourses, we see little engagement with the very serious limits posed by existing marginalisation, and the role that structural, political and economic interests play in privileging certain climate change narratives above others. We also see how certain voices are prioritised in official and semi-official spaces, exacerbating unequal power relations and patters of exclusion.

A core question for most of the authors here has been whether rights-based strategies can provide solutions or a stronger voice for the poor in the context of climate change. The chapters in this book suggest that we cannot yet fully answer this question, in part because rights- and justice-based strategies are not fully developed in relation to the problems posed by climate change. What we do see is avoidance – or merely rhetorical use – of rights, alongside an emergent but not yet fully developed attention to the relationship between socio-economic and environmental rights. We see climate justice falling through the cracks of technocratic, win-win language that does not engage the difficult issues. The avoidance of a focus on rights, justice and addressing the needs of the poor constitutes an evasion of responsibility. It also constitutes a major gap in climate change research, and one that must be filled.

This book, along with a Special Issue on 'Climate Change Justice: Narratives, Rights and the Poor' of the *South African Journal on Human Rights* ((2013) 29), would not have occurred without the financial support, which we gratefully acknowledge, from the Norwegian Research Council grant for the project on: Climate change discourse, rights, and the poor: Scientific knowledge, international political discourse, and local voices. We are also grateful to the Chr Michelsen Institute (CMI) and the Socio-Economic Rights Institute of South Africa (SERI) for providing the institutional support and space to pursue this project (and to CMI for awarding a visiting researcher grant to Jackie Dugard during June 2013 to finalise the book process). Finally, we are indebted to, and inspired by, all the authors for their commitment to this project, and for their dedication towards, as well as critical reflection about, climate change justice.

Jackie Dugard, Asunción Lera St. Clair and Siri Gloppen,
Norway, June 2013

TALK IS CHEAP

THE ROLE OF SOCIAL JUSTICE AND POVERTY IN SOUTH AFRICA'S NATIONAL CLIMATE CHANGE RESPONSE WHITE PAPER

KJERSTI FLØTTUM AND ØYVIND GJERSTAD

I INTRODUCTION

The challenges of climate change raise fundamental questions about social justice, equity and human rights. It is a global phenomenon but it is often experienced locally. Moreover, those who are least prepared and most likely to suffer first and in the most dramatic way from the consequences of climate change are those who have contributed the least to greenhouse gas (GHG) emissions leading to climate change.[1]

This issue is highly relevant in the case of South Africa, a developing country, which is both responsible for substantial GHG emissions and expected to suffer from the adverse effects of projected climate change. More importantly, the loss of life, health and property is expected to be exacerbated by the country's significant socio-economic inequalities, as marginalised population groups are least capable of adapting to the projected changes.[2] Climate change can thus be considered as a threat to social justice.[3] In this chapter, we will investigate to what extent issues related to social justice and poverty are emphasised in the National Climate Change Response White Paper published in 2011 by the government of South Africa. This document outlines 'the vision for an effective climate change response and the long-term transition to a climate-resilient, equitable and internationally competitive lower-carbon economy and society'.[4] More specifically, we will examine how the government takes into consideration the two orientations which have dominated both scholarly and political debate: the legal perspective which seeks to define the relevant rights in terms of national and international law and the ethical branch which argues for a broader approach by emphasising values, ethics and beliefs, providing for 'authoritative advocacy'.[5]

---

1   See K O'Brien, AL St. Clair & B Kristoffersen (eds) *Climate Change, Ethics and Human Security* (2010); AL St. Clair & D McNeill *Global Poverty, Ethics, and Human Rights: The Role of Multilateral Organisations* (2009).

2   P Bond, D Rehana & G Erion (eds) *Climate Change, Carbon Trading and Civil Society. Negative Returns on South African investments* (2009).

3   E Cameron 'Development, Climate Change and Human Rights. From the Margins to the Mainstream?' *World Bank Social Development Papers* 123 (2011) 1–3.

4   Republic of South Africa *National Climate Change Response White Paper* (2011) 10.

5   Cameron (note 3 above).

The seriousness of the issue is clearly stated in the *Human Development Report 2007/2008:*[6]

> Climate change confronts us with enormously complex questions that span science, economics and international relations. These questions have to be addressed through practical strategies. Yet it is important not to lose sight of the wider issues that are at stake. The real choice facing political leaders and people today is between universal human values, on the one side, and participating in the widespread and systematic violation of human rights on the other.

In the interpretation of the South-African White Paper, it is important to note that there has been a clear opposition to the current political regime responsible for the White Paper. Many voices have objected by claiming that the politicians in position tend to 'Talk Left, so as to Walk Right',[7] in particular with respect to the country's carbon-intensive economy. We will come back to this in part II, where we also bring in the process having led up to the White Paper.

With this as a backdrop, we see the following questions as relevant: To what extent is climate change presented as an issue of human rights and social justice in the National Climate Change Response White Paper? Is there an emphasis on the legal or the ethical dimension of this issue? In which contexts does the White Paper present rights issues and concerns of the poor as relevant? What is the role of the poor in the 'story' of climate change?

We will attempt to answer these questions through a combined lexical and narrative analysis, which will be related to pertinent contextual factors to the extent we find it justified. Our main methodological approach will then be linguistic and discursive. We believe it is crucial to go beyond content analysis and undertake a linguistically-based investigation, which can identify the different voices and actors that may be intertwined in this political discourse. We argue that a combination of two approaches will be relevant: a lexical analysis of selected keywords on the word/sentence level and a narrative analysis on the text level, supported by a polyphonic (or multivoiced) approach. These approaches allow us to go beyond a mere surface analysis by revealing implicit and covert rhetorical strategies. Our aim is to explain possible text interpretations which are not easily accessible, thereby offering scientific justifications to potential critical readings. More specifically the combined lexical-narrative approach can show to what extent the White Paper constructs climate change as a poverty issue, and whether it conceptualises this complex problem in an ethical or legal framework.

In part II we will first try to give a brief overview of the relevant political context in which the White Paper has been produced, as well as of the process

---

6   *Human Development Report 2007/2008. Fighting Climate Change: Human Solidarity in a Divided World.* Overview 2; See K Fløttum & T Dahl 'Different Contexts, Different "Stories"? A Linguistic Comparison of Two Development Reports on Climate Change' (2012) 32 *Language & Communication* 14; D Gasper, AV Portocarrero & AL St. Clair 'Climate Change and Development Framings: a Comparative Analysis of the Human Development Report 2007/8 and the World Development Report 2010' (2011) *Institute of Social Studies Working Paper* 528 <http://www.iss.nl/news_events/iss_news/detail/article/32244-wp-528-climate-change-and-development-framings-a-comparative-analysis-of-the-human-developmen/>.

7   P Bond *Talk Left, Walk Right. South Africa's Frustrated Global Reforms* 2 ed (2006).

leading from the Green Paper[8] up to the text of the White Paper. Then we will give a general overview of the structure and content of the White Paper. This part will constitute the context for the interpretation of our findings developed through the lexical and narrative analyses.

In part III, devoted to a lexical analysis, we present a quantitative overview of the following keywords: 'equality', 'equitable', 'equity', 'fair', 'inequalities', 'just', 'poor', 'poverty', 'right' and 'vulnerable'. We will then refine the keyword count by taking a closer look at their distribution in the text structure, in order to determine in what contexts of the proposed 'climate change response' rights and social justice are presented as salient factors. Another word that is relevant to political discourse on climate change is 'resilience', due to the increased focus on adaptation strategies. Even though this word is relatively frequent in the White Paper, we have chosen not to include it because its linguistic meaning is not directly related to social justice.

In part IV we will further elaborate on these findings within the framework of a text linguistic narrative theory[9] which is based on the hypothesis that discourses conceptualise political issues as plots, ie 'they recount some kind of problem or complication, followed by a sequence of events or actions which take place to achieve some particular effect(s)'.[10] This narrative analysis will help to determine whether and in what manner the issues of poverty and social justice play a part in the 'climate change story'. It will be further developed through a polyphonic perspective revealing different hidden voices and actors. Finally, the two approaches – lexical and narrative – will be combined in order to better answer the questions we raised above.

In our final remarks (part V), we will sum up and discuss our main findings, which can be formulated as follows: (1) the legal case for action is virtually absent; (2) ethical perspectives are modestly present in the argumentation of fairness; (3) the government is assuming the role of 'hero' of its own narrative; (4) the poor are presented as incapable agents and beneficiaries of mostly government action. We will close by addressing some possible paths for further research.

II CONTEXT AND CONTENT OF THE WHITE PAPER

A white paper serves to outline legislative intentions, which are developed through a process starting with the publication of a green paper by the government.[11]

---

8   Department of Environmental Affairs *National Climate Change Response Green Paper* (2010) <http://www.climateresponse.co.za/resources>.

9   See K Fløttum 'A Linguistic and Discursive View on Climate Change Discourse' (2010) 58 *La revue du GERAS, Asp* 19; K Fløttum 'Narratives in Reports about Climate Change' in M Gotti & CS Guinda (eds) *Narratives in Academic and Professional Genres* (in press); for a political approach, see MD Jones & MK McBeth 'A Narrative Policy Framework: Clear Enough to be Wrong?' (2010) 38 *Policy Studies J* 329.

10   K Fløttum & Ø Gjerstad 'Arguing for Climate Policy through the Linguistic Construction of Narratives and Voices: the Case of the South African Green Paper "National Climate Change Response"' 5 (submitted *Climatic Change)* now published in *Climatic Change* (2012) 115 (3-4), DOI: 10.1007/s 10584-012-0654-7.

11   For a linguistic discourse analysis of the 'National Climate Response Green Paper' see Fløttum & Gjerstad (note 10 above).

The green paper is subjected to critical review through a period of consultation with stakeholders, such as business and civil society organisations. As the end product of this process, the white paper is intended to include the insight and concerns of the various stakeholders. The following three subsections aim to give an overview of the main issues and controversies surrounding South Africa's climate strategy in general and the White Paper in particular. Our presentation does not constitute an exhaustive account of the issue but will provide some necessary context to the linguistic and textual analysis of parts III and IV.

When we in the following refer to the formal authors of the White Paper, ie the Government of the Republic of South Africa, we will use the abbreviation SAG; when referring to the Republic of South Africa, we will use SA.

(a) The political and institutional process leading up to the White Paper

The White Paper was the result of a seven-year process, which had started with the National Climate Change Response Strategy for South Africa,[12] a 48-page document published by the Department of Environmental Affairs and Tourism in 2004.[13] This publication was followed by the 'Long Term Mitigation Scenario' (LTMS) in 2008, the purpose of which was 'to outline different scenarios of mitigation action by South Africa, to inform long-term national policy and to provide a solid basis for our position in multi-lateral climate negotiations on a post-2012 climate regime'.[14] In 2010 the SAG released the National Climate Change Response Green Paper.[15] At the Gauteng workshop following this release, the Environmental Affairs director Joanne Yawitch stated that 'We as a country need to accept that our emissions will peak during 2020 – 2025, plateau up until 2035 and then only start to decline', in line with the LTMS.[16] Many non-governmental organisations (NGOs), activists and academics deemed this emissions scenario wholly insufficient. According to professor of political economy Patrick Bond, this long-term timetable was a ruse which paved the way for the 'multi-billion dollar financing decisions on Eskom coal-fired mega power plants',[17] while virtually ensuring human suffering from drought and other man-made natural disasters. According to

---

12   'National Climate Change Response Green Paper' (2011) 6(2) *25 Degrees in Africa* <http://www.25degrees.net/index.php/component/option,com_zine/Itemid,146/id,1244/view,article/>.

13   Department of Environmental Affairs and Tourism 'A National Climate Change Response Strategy for South Africa' (2004) <http://www.google.no/url?sa=t&rct=j&q=a%20national%20climate%20change%20response%20strategy%20for%20south%20africa&source=web&cd=1&cad=rja&ved=0CCcQFjAA&url=http%3A%2F%2Funfccc.int%2Ffiles%2Fmeetings%2Fseminar%2Fapplication%2Fpdf%2Fsem_sup3_south_africa.pdf&ei=fJ5NUMiGBZGk4ATf34CgBg&usg=AFQjCNEtxEd4Im-Yd4arBGt9rH4TygpLBw>.

14   Energy Research Centre 'Long Term Mitigation Scenarios: Technical Summary' (2007) 2 <http://www.erc.uct.ac.za/Research/publications/07ERC-LTMSTechnical_Summary.pdf>.

15   Department of Environmental Affairs (note 8 above).

16   National Climate Change Response Paper (note 12 above).

17   P Bond *Politics of Climate Justice: Paralysis Above, Movement Below* (2012) 151.

Bond, the Green Paper amounted to nothing more than 'greenwashing', as the SAG was once again 'talking left, so as to more rapidly walk right':[18]

> Consistent with Washington's irresponsible climate agenda, Pretoria's Green Paper suggests we 'limit the average global temperature increase to at least below 2°C above pre-industrial levels', yet this target is so weak that scientists predict nine out of ten African farmers will lose their ability to grow crops by the end of the century. In contrast, the 2010 Cochabamba People's Agreement demanded no more than a 1–1.5°C rise. This is a vast difference when it comes to emissions cuts needed to reach back from present levels of close to 400 parts per million of CO_2 equivalents in our atmosphere to the 300 parts per million that science requires and that the Cochabamba conference demanded.[19]

This view was shared by among others Earthlife Africa, who considered that a temperature rise 'below 2° could still have potentially catastrophic impacts on South Africa in the medium to long term', a scenario suggested by the Green Paper itself. Earthlife Africa also pointed to inconsistencies and imprecisions in the data used in the Green Paper.[20]

(b) Structure and content of the White Paper

The question is how the SAG took such reservations into account in the writing of the White Paper (WP). The structure of the document is very different from the Green Paper (GP), a fact noted by environmental NGOs in the subsequent hearings in late 2011.[21] While the GP structures its 'Policy approaches and actions' (chapter 5) according to sectors (for example water, energy, human health, etc), the WP makes a primary distinction between 'Adaptation' (chapter 5) and 'Mitigation' (chapter 6). Furthermore, the WP presents eight 'near-term flagship programmes', which are designed to address challenges such as energy, water and transport, and which are completely absent from the GP.

The WP is divided into 13 chapters, in addition to an initial 'Executive summary', which presents its main contents. The Introduction explains the phenomenon of climate change, and outlines the SAG's on-going and future efforts to mitigate and adapt to the projected changes, in a national as well as international context. The second chapter, 'National climate change response objective', gives a short presentation of SA's dual challenge of adaptation and mitigation. Chapter 3 presents the nine 'Principles' which guide this objective, including 'Common but differentiated responsibilities and respective capabilities', 'Equity', and 'Uplifting the poor and vulnerable'. Chapter 4 outlines the 'National climate change response strategy' in bullet points, describing a decision-making process based on experience, costs and

---

18 Ibid 151–2.
19 Ibid 153.
20 Earthlife Africa 'Earthlife Africa Jhb Submission on National Climate Change Response Green Paper 2010' (2011) 2 <http://www.google.no/url?sa=t&rct=j&q=earthlife%20africa%20jhb%20submission%20on%20national%20climate%20change%20response%20green%20paper%202010%20&source=web&cd=1&cad=rja&ved=0CCIQFjAA&url=http%3A%2F%2Fd2zmx6mlqh7g3a.cloudfront.net%2Fcdn%2Ffarfuture%2Ftime%3A1299749293%2Ffiles%2Fdocs%2F110308earthlife_0.doc&ei=o6lNUKOTGfHQ4QShj4BI&usg=AFQjCNGfrv0j3LVrC_qqbcjzbZgY9MHD-A>.
21 Parliamentary Monitoring Group 'White Paper on Climate Change Public Hearings' (2011) <http://www.pmg.org.za/report/20111101-public-hearings-national-climate-change-white-paper-2011>.

benefits, risks and incentives and disincentives for behaviour change, among other factors. Notably the SAG retains the emissions scenario from 2008, 'a benchmark "peak, plateau and decline" GHG emission trajectory where GHG emissions peak between 2020 and 2025, plateau for approximately a decade and begin declining in absolute terms thereafter'.[22] Chapter 5 discusses adaptation measures for several sectors, including water, agriculture, biodiversity, health, human settlements and disaster risk management. Chapter 6 presents mitigation efforts, underlining from the outset that the proposed peak-plateau-decline trajectory is not an absolute commitment but depends on the mobilisation of financial resources by developed countries:

> As a responsible global citizen and as a global citizen with moral as well as legal obligations under the UNFCCC and its Kyoto Protocol, South Africa is committed to contributing its fair share to global GHG mitigation efforts in order to keep the temperature increase well below 2°C. In this regard, on 6 December 2009, the President announced that South Africa will implement mitigation actions that will collectively result in a 34% and a 42% deviation below its 'Business As Usual' emissions growth trajectory by 2020 and 2025 respectively. In accordance with Article 4.7 of the UNFCCC, the extent to which this outcome can be achieved depends on the extent to which developed countries meet their commitment to provide financial, capacity-building, technology development and technology transfer support to developing countries. With financial, technology and capacity-building support, this level of effort will enable South Africa's GHG emissions to peak between 2020 and 2025, plateau for approximately a decade and decline in absolute terms thereafter.[23]

Chapter 7, 'Managing response measures', addresses the issue of potential negative economic impacts of measures, pertaining especially to mitigation. Chapter 8 presents the 'Near-term priority flagship programmes' for the near-term. These include public works, water conservation and demand management, renewable energy, energy efficiency and energy demand, transport, waste management, carbon capture and sequestration, and lastly adaptation research. Chapters 9 and 10 discuss respectively job creation and the organisation of public institutions with regard to the projected socio-economic changes. The mobilisation of resources is outlined in chapter 11, while the measurement and monitoring of climate change and response efforts are covered in chapter 12. Lastly, the 'Conclusion' briefly summarises the challenges represented by climate change, and restates the SAG's commitment to put in place an effective response.

(c)   Reactions to the White Paper

The reactions to the WP merit a comprehensive study in and of themselves. In the present chapter, we limit our account to some of the responses put forward by the NGOs Earthlife Africa, Greenpeace and World Wildlife Fund for Nature (WWF), presented in writing and at a parliamentary hearing on 31 October 2011. While the responses of these organisations may not represent civil society as a whole, they permit us to identify some of the most contentious aspects of the WP.

---

22   *White Paper* (note 4 above) 13.
23   Ibid 24–5.

In its comments submitted to parliament, the WWF was 'greatly encouraged by the National Climate Change Response White Paper',[24] believing that the document would 'contribute to South Africa's international prestige as host of COP17'.[25] The introduction of the response is characterised by praise: 'The Department of Environmental Affairs and other participant arms of government are to be congratulated on driving this outcome ...'. More specifically the WWF mentioned the carbon budget approach, including mandatory mitigation plans involving a carbon tax, as a step in the right direction. This NGO also considered the Flagship Programmes as 'well-conceived'.[26]

Earthlife Africa was less laudatory in its response to the WP, but stated that the WP was 'much improved' over the GP: 'Notably the exclusion of nuclear power from the document, the addition of numerical values for mitigation targets, a carbon budget approach, and a commitment to keep, "well below a maximum of 2°C above pre-industrial levels"'.[27] Greenpeace Africa also welcomed the exclusion of nuclear power, as well as the inclusion of mitigation targets.[28]

However, there was dissatisfaction among several NGOs, at both the process of public consultation and at the proposed policies. Regarding the procedural grievances, the SA parliament had given notice of only eight days between the release of the WP and the hearing during which Greenpeace Africa, Earthlife Africa Johannesburg, the Environmental Monitoring Group and the WWF were invited to present their reactions. In addition, Greenpeace considered that there seemed to be 'an *element of haste*' in the final stages of the drafting of the WP (original emphasis),[29] which they attributed to the imperative of publishing it ahead of COP17.[30] In other words, this 'hasty' publication could be interpreted as a showcase, displayed to the advantage of SA as the host of COP17.

As for the content of the WP, the NGOs considered that the future peak-plateau-decline trajectory needed to give way to an immediate emissions peak, followed by a decline. There was also uncertainty about the baseline in relation to emissions, as well as the trajectory range and the nature of the flagship programmes developed in the WP. Other concerns included the SAG's reliance on future carbon capture and storage technology, as such methods had not yet proven to be viable. Renewable energy development was seen to be a much more promising area for investment. Furthermore, the NGOs considered that revenues from a carbon tax needed to be 'ring-fenced' so as

---

24 WWF South Africa 'Comments by WWF South Africa in response to the National Climate Change Response White Paper' (2011) 8 <http://d2zmx6mlqh7g3a.cloudfront.net/cdn/farfuture/mtime:1320400246/files/docs/111101wwf-edit.pdf>.

25 Ibid 1–2.

26 Ibid 1.

27 Earthlife Africa Jhb 'Comments on the National Climate Change Response White Paper' (2011) 1 <http://d2zmx6mlqh7g3a.cloudfront.net/cdn/farfuture/mtime:1320400116/files/docs/111101earthlife.pdf>.

28 Greenpeace Africa 'GPAfr Comments on Climate Change White Paper' (2011) 4 <http://d2zmx6mlqh7g3a.cloudfront.net/cdn/farfuture/mtime:1320308704/files/docs/111101gpa_0.ppt>.

29 Ibid 3.

30 PMG (note 21 above).

not to be diverted from renewable energy development. Earthlife Africa also feared that companies such as Synthetic Oil Limited and Eskom would be able to pass the costs of carbon taxes on to consumers, a problem the SAG needed to address through legislation. Finally in its written response to the WP, Earthlife Africa pointed to the legal imperative of emissions reductions, as the right to a clean and healthy environment was unconditional:

> Parliament, under the Bill of Rights' Section 24 of the Constitution, is specifically mandated to ensure that the citizens of this country have a clean, safe and healthy environment. This should be the primary rule under which Parliament engages with the White Paper. Accordingly, Parliament cannot but overrule, reject, adapt, amend, or otherwise alter the White Paper if the Constitutional rights of the citizens of South Africa are likely to be infringed upon by the White Paper. A country that is blighted by the consequences of unmitigated global warming will not be a clean, safe or healthy environment for South Africans to live in; such a situation would be a complete and utter violation of the Constitution and an abject failure of all organs of the state to rule in accordance with its social contract with the body politic.[31]
>
> The last point is the conditional nature of the mitigation trajectory in the White Paper, namely dependent upon international finance. This is contrary to the Constitution. Our rights are not dependent on ODA, carbon finance, or donor funding. They are inalienable. South Africa is required to adopt emissions reduction even if no money comes from international finance, according to its Constitution.[32]

At the hearing Tristen Taylor of Earthlife Africa reiterated the importance of constitutional rights, to which chairperson Johnny de Lange replied that such rights could not be considered as absolute, because s 36 of the Constitution of the Republic of South Africa, 1996 limited rights in the event that two or more rights clashed or contradicted each-other.[33] The question is to what extent rights, and particularly the rights of the poor, are given a place in the WP, beyond their submission to financial conditions, as identified by Earthlife Africa. This is one of the main questions we will investigate in this chapter.

III   POVERTY, EQUITY AND RIGHTS IN THE WHITE PAPER – LEXICAL ANALYSIS

The purposes of the lexical analysis to be undertaken in this part can be explained as follows. First, the counting of keywords pertaining to a specific topic allows us to determine the salience of that topic in the given text. Second, by examining the distribution of keywords we get an indication of the topic's salience in different contexts, for example in mitigation versus adaptation. Third, the selected keywords which will be examined here pertain to either legal or ethical dimensions of climate change. Thus, differences in frequency can indicate which of these dimensions is most important in the WP.

The lexical analysis will focus on a selection of keywords that we consider as relevant to the issue of rights and social justice. We have not counted keyword occurrences which do not pertain to the issue of rights and social justice within

---

31   Earthlife Africa Jhb (note 27 above) 2.
32   Ibid 10.
33   PMG (note 21 above).

SA,[34] nor those which represent different meanings due to homonymy and polysemy.[35] Furthermore, regarding the 'poor', we have counted occurrences of both the noun and the adjective, including 'poorest communities'.[36] Finally, we have counted only the occurrences of 'vulnerable' which characterise population groups or individuals, not, for example, economic sectors. The results are listed in Table 1.

Table 1: Keywords

Key words	equal-ity	equi-table	equity	fair	inequali-ties	just	poor	pov-erty	right	vulner-able
Number	1	5	1	4	2	2	17	11	3	23

We see that words pertaining to poverty and vulnerability ('poor', 'poverty', 'vulnerable') outweigh those that concern fairness and legality ('equality', 'equitable', 'equity', 'fair', 'inequalities', 'just', 'right') by almost three to one – 51 to 18 to be precise. In that respect it is interesting to note that the word 'justice' is absent from the White Paper. We will discuss in further detail this discrepancy and how it indicates the SAG's choice of arguments when discussing the problem of poverty within the larger issue of climate change.

Before having a closer look at examples representing the different ways in which the keywords are used to create and argue for policies, let us have a look (Table 2) at the manner in which they are distributed in the text.

---

34   For example, in the case of 'vulnerable', we have left out occurrences such as the following: 'South Africa may be economically vulnerable to measures taken both internationally and nationally, to reduce GHG emissions.'

35   For example 'extremely *poor* air quality conditions' *White Paper* (note 4 above) 18.

36   Ibid 21.

Table 2: Distribution of keywords

Keywords / Chapters	equality	equitable	equity	fair	inequalities	just	poor	poverty	right	vulnerable	Sum	Total N of words	Relative frequency of keywords
Executive summary	0	0	0	0	0	1	0	0	1	0	2	1428	0.0014
1. Introduction	0	1	0	0	0	0	1	2	0	1	5	1894	0.0026
2. National climate change response objective	1	1	0	0	0	0	0	1	0	0	3	134	0.0224
3. Principles	0	0	1	1	1	0	3	0	1	4	11	332	0.0331
4. The South African climate change response strategy	0	1	0	0	0	0	0	2	0	1	4	1391	0.0029
5. Adaptation	0	0	0	1	1	0	8	2	1	11	24	5454	0.0044
6. Mitigation	0	0	0	0	0	0	1	1	0	0	2	2866	0.0007
7. Managing response measures	0	1	0	0	0	1	0	0	0	0	2	282	0.0071
8. Near-term priority flagship programmes	0	0	0	0	0	0	0	2	0	0	2	1304	0.0015
9. Job creation	0	0	0	0	0	0	0	1	0	2	3	892	0.0034

Keywords / Chapters	equality	equitable	equity	fair	inequalities	just	poor	poverty	right	vulnerable	Sum	Total N of words	Relative frequency of keywords
10. Mainstreaming climate-resilient development	0	1	0	0	0	0	4	0	0	4	9	3845	0.0023
11. Resource mobilisation	0	0	0	1	0	0	0	0	0	0	1	2285	0.0004
12. Monitoring and evaluation	0	0	0	1	0	0	0	0	0	0	1	1050	0.0010
13. Conclusion	0	0	0	0	0	0	0	0	0	0	0	295	0

We see that the issue of poverty, rights and social justice is present in all chapters, except for the 'Conclusion' (chapter 13) – which is an important finding in itself. Interestingly, it is much more prevalent in the chapter titled 'Adaptation' (chapter 5) than in the one titled 'Mitigation' (chapter 6), with a relative keyword frequency which is more than six times higher. Through the following examples we will look more closely at the WP's conceptualisation of poverty in relation to adaptation and mitigation, respectively. Let us first look at two examples from chapter 5, on climate change adaptation (our emphasis in bold in the examples):

(1)   All states in the Southern African sub-region face the challenges of rural and urban **poverty**, limited water or access to water resources, food insecurity, and other development challenges. Thus, although countries of the sub-region may have differing developmental priorities, they often face similar risks due to climate change and may also have similar adaptation needs. South Africa will therefore strive to develop climate change adaptation strategies based on risk and vulnerability reduction, in collaboration with its neighbours where appropriate, and seek to share resources, technology and learning to coordinate a regional response. A regional approach that achieves climate resilience will have significant socio-economic benefits for South Africa, including a smaller risk of unmanaged regional migration.[37]

(2)   In the medium to long term, the *Water for Growth and Development Framework,* which has a 2030 planning horizon, aims to balance the critical role of water in terms of both **poverty alleviation** (ensuring **the constitutional right** to a reliable and safe water supply) and economic development (be it for domestic, industry, mining, agricultural or forestry use). Water vulnerability and response must also be adequately factored into this framework document.[38]

As these excerpts demonstrate, adaptation is represented as an integral part of poverty alleviation and, in the case of (2), as imposed by constitutional constraints. Let us have a look at the two cases found in chapter 6 on mitigation, starting with the first paragraph:

(3)   South Africa's approach to mitigation is informed by two contexts: first, its contribution as a responsible global citizen to the international effort to curb global emissions; and second, its successful management of the development and **poverty eradication challenges** it faces. The National Climate Change Response is intended to promote adaptation and mitigation measures that will make development more sustainable, both in socio-economic and environmental terms.[39]

(4)   This policy identifies or sets up processes to identify the optimal combination of actions sufficient to meet the National Climate Change

---

37   Ibid 14.
38   Ibid 16.
39   Ibid 24.

Response Objective. Factors to be considered include not only the mitigation potential, the incremental and direct cost of measures, but also the broader impact on socio-economic development indicators (such as employment and income distribution), our international competitiveness, **the cost to poor households** and any negative consequences for key economic sectors.[40]

While (3) only hints at the potential conflict between mitigation and poverty alleviation, this is more explicitly formulated in (4), where there is a clear tension between SA's mitigation responsibilities and the needs of the poor. This is further emphasised in chapter 10 ('Mainstreaming climate-resilient development'), on the development of a carbon tax policy:

(5)   ... measures will be taken, either in tax design or through complementary expenditure programmes, to offset the burden such a tax will place on **poor households.**[41]

These five examples serve to illustrate the SAG's dual conceptualisation of poverty eradication as both concomitant to and in conflict with climate change policies. In the WP we have identified nine sentences in which poverty alleviation and mitigation are presented as competing priorities. Four of these sentences do not discuss measures to help the poor, mentioning instead poverty as an explicit or implicit justification for limited reduction targets in the context of international negotiations. This is in line with the SAG's discourse in previous climate change negotiations, which Patrick Bond qualifies as a 'ruse':

> After Van Schalkwyk came Sonjica as environment minister, and then in late 2010, her replacement Edna Molewa. In Cancún, Molewa slowed progress on getting binding emissions cuts, because '[w]e believe that it is quite important that as developing countries we also get an opportunity to allow development to happen because of poverty. We need to allow space for us to actually introduce those emissions [reductions] over time, because developed countries have gone through the processes.' The ruse here was that South Africa's extremely high emissions contributed to poverty-reduction, rather than the opposite, as a result of big business over-consuming scarce electricity and driving up the price beyond poor people's ability to pay.[42]

As the energy costs are currently passed on to the poor, to the benefit of large industrial enterprises, the SAG's argument that the imperative of poverty alleviation stands in the way of mitigation reductions appears simplified at best.

As for the argumentative basis for protecting the least privileged population groups of SA, we can make a distinction between three main categories among the examples found in the White Paper: (i) poverty alleviation and the protection of vulnerable groups as an argument on its own terms, with no further reference to an ethical or legal principle; (ii) the argumentation of fairness, in which the ethical principle is explicit; (iii) the legal argumentation, in which efforts are presented as not only desirable or necessary but obligatory.

---

40   Ibid 27.
41   Ibid 40.
42   Bond (note 17 above).

The first two categories could be said to correspond to the trend of ethical argumentation in favour of human rights, as described by Cameron,[43] whereas the third category represents the legal branch of human rights.[44] The following examples represent the three categories:

(6) In both the agriculture and commercial forestry sectors synergy and overlap exists between adaptation and mitigation measures, and climate-resilient sectoral plans have the potential to directly address the plight of those most impacted by climate change – **the rural poor.**[45]

(7) At the national level, the challenge will be to effectively manage and reduce the economic risks, and build on and optimise the potential opportunities, and to ensure a smooth and just transition to a climate-resilient, **equitable** and internationally competitive lower-carbon economy and society.[46]

(8) This response is guided by principles set out in the Constitution, **the Bill of Rights**, the National Environmental Management Act, the Millennium Declaration and the United Nations Framework Convention on climate change.[47]

In (6) there is no mention of rights or principles regarding poverty, but the prospect of relieving the plight of the poor through political measures serves as a sufficient argument to initiate those policies. In (7) equity is presented as one of the overarching goals of the SAG's climate policies, and constitutes an example of category (ii). The last of the three examples lays out the legal framework for action, including the *Bill of Rights*, which protects the rights of all South Africans, thus belonging to category (iii).

The first two examples could both be said to evoke the ethical dimension of climate change policies. Even though (6) does not explicitly refer to a moral principle, the reference to indigence carries moral weight much in the same way as the mention of the principle of equity in (7). This is in contrast to (8), in which the government underlines its own legal obligations with regard to the citizens of South Africa. By doing so it imposes not only a responsibility but also a legal constraint on itself.

The question is how prevalent one category is compared to the others, focusing on the distinction between categories (i) and (ii) on the one hand and category (iii) on the other. In order to give an indication we have counted the number of sentences representing the different categories, on the basis of the keywords we have chosen. Category (i) includes sentences containing one or more of the words 'poor', 'poverty' and 'vulnerable', but none of the other words. Sentences belonging to category (ii) includes one or more of the following words: 'equality', 'equitable', 'equity', 'fair', 'inequalities' and

---

43   Cameron (note 3 above).
44   Ibid.
45   *White Paper* (note 4 above) 17.
46   Ibid 8.
47   Ibid 5.

'just'; whereas category (iii) includes the word 'right'. This gives the sentence count as shown in Table 3.

Table 3: Number of sentences in the argumentation categories

Category	(i)	(ii)	(iii)
No of sentences	41	11	3

This means 74.5 per cent of the sentences counted belong to category (i), which shows that the argumentation is based less on explicit principles and rights than on the connotations of the words 'poor', 'poverty' and 'vulnerable'. More importantly, only 5.5 per cent of the sentences include a reference to legal rights. The three examples are listed here:

(9)   In the medium to long term, the Water for Growth and Development Framework, which has a 2030 planning horizon, aims to balance the critical role of water in terms of both **poverty alleviation** (ensuring **the constitutional right** to a reliable and safe water supply) and economic development (be it for domestic, industry, mining, agricultural or forestry use). Water vulnerability and response must also be adequately factored into this framework document.[48]

(10) This response is guided by principles set out in the Constitution, **the Bill of Rights**, the National Environmental Management Act, the Millennium Declaration and the United Nations Framework Convention on Climate Change.[49]

(11) The achievement of South Africa's climate change response objective is guided by the principles set out in the Constitution, **the Bill of Rights**, the National Environmental Management Act (NEMA), the MDGs and the UNFCCC.[50]

Only (9) includes a concrete reference to a legal right pertaining to the issue of climate change, while (10) and (11) more broadly mention the principles set out in the Bill of Rights as a guide to the climate change response strategy. However, (11) is followed by a more concrete set of principles, among which we find the following (original emphasis):

(12) *   **Equity** – ensuring a fair allocation of effort, cost and benefits in the context of the need to address disproportionate vulnerabilities, responsibilities, capabilities, disparities and inequalities.

    *   **Special needs and circumstances** – considering the special needs and circumstances of localities and people that are particularly vulnerable to the adverse effects of climate change, including vulnerable groups such as women, and especially poor and/ or rural women; children, especially infants and child-headed families; the aged; the sick; and the physically challenged.

---

48   Ibid 16.
49   Ibid 5.
50   Ibid 11.

*   **Uplifting the poor and vulnerable** – climate change policies and
         measures should address the needs of the poor and vulnerable
         and ensure human dignity, whilst endeavouring to attain
         environmental, social and economic sustainability.

While these bullet points clearly state the need for equity, poverty alleviation
and the protection of vulnerable groups, the listed principles are not directly
linked to legal constraints on the government. In other words there is no
explicit reference to any rights enshrined in law, only to principles which
have multiple sources, some of which do not have legal status, such as the
Millennium Development Goals (MDGs) and the United Nations Framework
Convention on Climate Change (UNFCCC). This leads to the conclusion that
the SAG presents only one example of its legal responsibilities to its citizens
in relation to climate change policies, namely 'the constitutional right to a
reliable and safe water supply' (example (9)).

In part IV we will move to a qualitative analysis of the WP's construction of
climate change as a narrative, a 'story' with a plot and different actors. At the
end of this analysis we will present an overview of keyword occurrences in
relation to the narrative. By combining the two approaches we can determine
which of our keywords are prevalent in the construction of the different stages
of the story on climate change.

IV   NARRATIVE ANALYSIS

In the lexical analysis above, we have seen that climate change is linked in a
very modest way to the legal rights of the poor. We have also observed that
concerns for the poor are more tightly related to adaptation than to mitigation.
In this part, we will develop our analysis within a narrative framework.
After a short presentation of the narrative model that we take as our point
of departure, we will use this theoretical perspective to identify the 'plot'
underpinning the 'climate change story' presented by the SAG, to examine to
what extent issues of rights and poverty are foregrounded, as well as to further
determine the roles played by the poor.

This macro-structural analysis will be strengthened by some comments of
linguistic features conveying polyphony or 'multivoicedness'.[51] A number of
linguistic markers, such as negation, connectives ('thus', 'but', 'however' etc)
and reported speech, give rise to voices other than that of the speaker at the
moment of utterance. The following example illustrates polyphony through
polemic negation. In July of 2012 the climate researcher Bill McGuire wrote
a blog post in *The Guardian*, titled 'Climate change is not science fiction,
Jeremy Clarkson'.[52] The negation in the title signals the presence of another

---

51   See Fløttum (note 9 above); K Fløttum & T Dahl 'Climate Change Discourse: Scientific Claims
      in a Policy Setting' (2011) 3 *Fachsprache* 205; Ø Gjerstad '*La polyphonie discursive. Pour un
      dialogisme ancré dans la langue et dans l'interaction*' (2011) doctoral dissertation; H Nølke, K
      Fløttum & C Norén *La ScaPoLine. La théorie scandinave de la polyphonie linguistique* (2004).
52   B McGuire 'Climate Change is not Science Fiction, Jeremy Clarkson' *The Guardian* (10 July 2012)
      <http://www.guardian.co.uk/environment/blog/2012/jul/10/climate-change-science
      -fiction-jeremy-clarkson>.

voice which is of the opinion that 'climate change is science fiction', and which is then identified as television host Jeremy Clarkson.

We see polyphony as particularly relevant since political narratives typically construct and reproduce patterns of interests and conflicts between different actors, which may be signalled through specific linguistic markers. From our exploratory analysis,[53] it is clear that there are numerous competing interests, implicitly or explicitly present, which the SAG has to take into consideration.[54] The polyphonic perspective will help us gain a deeper understanding of how these are represented and integrated in the text.

(a) Narrative structure

Preliminary studies of various documents related to climate change indicate that the narrative structure may be fruitful as an overarching frame of understanding these documents which then can be considered to be part of what we call *climate change narratives*.[55] By 'climate change narrative' we refer to text and talk that present climate change as a certain type of problem, with implicit or explicit suggestions for action. The notion of narrative has often been used in a rather non-critical way. Here we want to return to the classical narrative structure (studied mostly in literary or fiction contexts), updated and developed by Adam[56] in five components: (1) Initial situation (or orientation); (2) Complication (creating difficulties; 'déclencheur' – release mechanism); (3) (Re-)action(s); (4) Resolution (the act of resolving; 'dénouement'); (5) Final situation.

In addition to these components, there may also be a moral or evaluation added to the narrative. Below we have constructed, for illustrative purposes, a short and very simple example of the macro-structure of a possible narrative sequence:

(13) 1. CO_2 emissions increased dramatically between 1990 and 2007.
2.   Global warming has caused serious problems in numerous regions.
3.   The UN organised an international summit in Copenhagen in 2009 (COP15) to discuss action on climate change.
4.   But the negotiating countries did not reach any binding agreement of measures to undertake.

---

53   See Fløttum & Gjerstad (note 10 above).
54   For more on different interests and voices in the climate change debate, see M Boykoff *Who Speaks for the Climate? Making Sense of Media Reporting on Climate Change* (2011); A Giddens *The Politics of Climate Change* (2009); H Haddad, T Morton & A Rabinovic 'The Roles of "Tone of Voice" and Uncertainty when Communicating Climate Change Information' (forthcoming) Presented at the British Psychological Society Social Section Conference Cambridge (September 2011); M Hulme *Why We Disagree About Climate Change* (2009); E Malone *Debating Climate Change* (2009); J Painter *Poles Apart: the International Reporting of Climate Scepticism* (2011); E Zaccai, F Gemenne & JM Decroly *Controverses climatiques, sciences et politique* (2012).
55   Fløttum (note 9 above).
56   JM Adam *La linguistique textuelle. Introduction à l'analyse textuelle des discours* 2 ed (2008).

5.   Climate change constitutes a serious threat to the planet, and those who have contributed least to the problems are the ones most vulnerable to the consequences.

All five components are not obligatory for the construction of a narrative. The core is constituted by the three middle components, and among them, the 'complication' (component 2) is mandatory. Our working hypothesis is that in climate change narratives, the complication factor is typically climate change itself. Through previous studies,[57] we see that the complication component can also give rise to new complications, new stories (for example on economic, technological, political, social or ethical development). This is illustrated in Figure 1, where each box represents a stage of a common narrative of emission cuts:

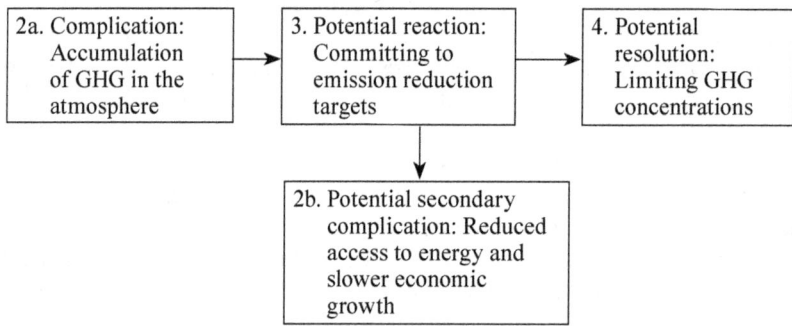

Figure 1: Climate change mitigation narrative

The goal of climate change mitigation is compromised by the expected socio-economic consequences (box 2b) of the necessary policies. How this secondary complication may be developed and how new 'stories' are constructed and framed are crucial questions for linguistic and discursive studies.[58]

Components 1, 3 and 4 can be realised at different degrees according to the situations, institutions, actors and voices involved. Component 5 may be present, for example, in descriptions of how our world will develop if we undertake such and such action or how our world will develop if we do nothing about the stated climate change complication. In addition, the place given to evaluative aspects is also relevant; ethical aspects are often integrated in climate change narratives. In the above example (13), the last part of the final situation could be interpreted as an evaluation ('those who have contributed least to the problems are the ones most vulnerable to the consequences').

Climate change narratives are of course not identical to classical narratives such as, for example, fairy tales (in the fictional world). However, there are certain crucial points that may be regarded as similar. In addition to the

---

57   Fløttum & Dahl (note 6 above).
58   For the question of 'framing', see MC Nisbet 'Communicating Climate Change: Why Frames Matter for Public Engagement' (2009) *Environment. Science and Policy for Sustainable Development* <http://www.environmentmagazine.org/Archives/Back%20Issues/March-April% 202009/Nisbet-full.html>.

typical trait of temporally ordering of events, a more important trait is the overarching characteristic of narratives as having a *plot*, ie they recount some kind of problem or complication, followed by a sequence of events and actions that take place to achieve some particular effect(s). A final trait that we will pay particular attention to here is the presence of actors: hero(es), villain(s) and victim(s).[59]

To sum up: the narrative has a plot to the extent that it recounts some kind of problem or *complication* (related to climate change), with different kinds of actors (humans, nature, society, countries); it may be followed by a sequence of events or *re-actions*, or more often (explicit or implicit) recommendations or imperatives of actions, which take place or should take place to achieve some particular effect(s) or a *final situation*; and, according to different interpretations of the complication factor, *ethical perspectives* may be advanced.

(b)   Plot, actors and the role of the poor in the White Paper

Already in the Green Paper on the country's climate change response, published in 2010,[60] the SAG characterised South Africa as follows, attributing a double role of villain and victim to the country (as a contributor to global climate change):

(14) South Africa is both a contributor to, and potential victim of, global climate change given that it has an energy-intensive, fossil-fuel powered economy and is also highly vulnerable to the impacts of climate variability and change.[61]

This attribution of the double role of victim and villain is clearly repeated in the White Paper:

(15) It is acknowledged that Africa, as a whole, has contributed least to GHG concentrations in the atmosphere, **but** also faces some of the worst consequences and generally has the least capacity to cope with climate change impacts. **However**, it is also recognised that South Africa is a relatively significant contributor to global climate change with significant GHG emission levels from its energy-intensive, fossil-fuel powered economy. **On the other hand**, South Africa is extremely vulnerable and exposed to the impacts of climate change due to our socio-economic and environmental context. Climate variability, including the increased frequency and intensity of extreme weather events, will disproportionately affect **the poor**.[62]

---

59   For a discussion of these in media studies, see E Eide, R Kunelius & V Kumpu (eds) *Global Climate – Local Journalisms. A Transnational Study of How Media Make Sense of Climate Summits* (2010).
60   *Green Paper* (note 8 above).
61   Ibid 4.
62   *White Paper* (note 4 above) 8–9.

This polyphonic, or 'multivoiced', passage is also demonstrating the many implicit voices that the SAG takes into account. The polyphonic markers 'but', 'however' and 'on the other hand' signal a series of concessions, where the final argument introduced by 'on the other hand' wins – 'South Africa is extremely vulnerable ...'. There is no doubt that between the voices pointing at SA as villain and as victim, the SAG identifies with and makes more prominent the ones according to which SA is presented as a victim, thus implicitly blaming the rich countries having contributed the most to the damaging GHG concentrations. We also notice that the poor are explicitly mentioned here, but without any further integration of rights issues. To sum up: the polyphonic perspective explains the manner in which connectives like 'but', 'however' and 'on the other hand' can be used as an effective rhetorical tool, as these polyphonic markers let the authors admit to the truth of points of view which may belong to others or to themselves in an internal discussion, before presenting the final point of view as being the most salient.[63] These connectives create an impression of attention to alternative views, while establishing a hierarchy that is beneficial to the argumentative purpose of the authors.

Furthermore, the above extract summarises the main plot of the SAG story. Climate change constitutes the main complication, representing a major threat to the country (emphasised through a thorough explanation in the 'Introduction' of the key findings of the IPCC and thus science-based knowledge); however, SA wants to contribute on a global level:

(16) ... as a responsible global citizen, and in keeping with its developing country status, its capability and its share of responsibility for the problem, South Africa is committed to making a fair contribution to the global effort to reduce GHG emissions.[64]

But their main story consists of recommendations or imperatives of actions, formulated in a general way as follows:

(17) South Africa will have to adapt to these impacts [of climate change] by managing its climate and weather-related risks, reducing its vulnerability and increasing the resilience of our society and economy to the adverse effects of climate change and variability;[65]

These actions should take place to achieve some particular effect(s), stated as follows in the 'Executive summary':[66] '[a] transition to a climate-resilient and lower-carbon economy and society'.

A simplified summarisation of the overall narrative represented in the White Paper can be formulated as follows: The *initial situation* and *complication*, represented by a description of South-Africa's specific natural and economic context and climate change respectively, are mainly developed

63   Nølke et al (note 51 above) 92–6.
64   *White Paper* (note 4 above) 19.
65   Ibid 10.
66   Ibid 5.

in the WP's 'Introduction', supported by objectives, principles and strategy in chapters 2 to 4 (also including some re-action measures). Then the major component in narrative terms is the re-action component, developed mainly in chapters 5 and 6, focusing on adaptation and mitigation respectively (with further elaborations in chapters 7 to 12). What is lacking if we were to have a 'full story' is an explicit *resolution* or *final situation* component (see also (c) below). There are, however, a few relevant expressions: 'a better life for all' ('Introduction'),[67] 'Creating a better South Africa and Contributing to a Better (and Safer) Africa and a better World' (Outcome 11),[68] and in the 'Conclusion':

(18) ... responding to climate change is a cross-generational challenge. The effects of action or inaction will not be felt immediately, but will have **significant consequences for future generations.**[69]

The many uncertainties related to the consequences of climate change as well as to the concrete realisation of the numerous measures to be undertaken can explain this lack of a clear vision of the future in 'a climate-resilient and lower-carbon economy and society' ('Executive summary').[70]

More importantly, the narrative analysis reveals that issues of human rights and social justice are given only a minor role in the plot. Both the *complication* and the *re-action* components emphasise the many challenges that SA is facing and will have to face in the future, especially of an economic nature, but without any clear address to the specific challenges that climate change will represent for ensuring the rights of the poor and social justice. We have already concluded that the legal case for action is virtually absent, save for one reference to the 'constitutional right to a reliable and safe water supply'.[71] What we are left with are the 11 sentences belonging to category (ii), as presented in Table 3. These sentences include at least one occurrence of 'equality', 'equitable', 'equity', 'fair', 'inequalities' or 'just', such as in the following example, which formulates one of the nine principles laid out in s III of the White Paper (original emphasis):

(19) **Equity** – ensuring a fair allocation of effort, cost and benefits in the context of the need to address disproportionate vulnerabilities, responsibilities, capabilities, disparities and inequalities.

Regarding the place given to the poor, we observe that they are present in the 'Adaptation' chapter and, to a certain extent, in the 'Mitigation' chapter.

We interpret these two chapters as sub-narratives within the larger overall narrative described above. This division into two chapters goes against the SAG's own conceptualisation of mitigation and adaptation: '[c]ategorising responses as either mitigation or adaptation responses can obscure the real and potential positive combined impact of these responses'.[72] However, the

67   Ibid 10.
68   Ibid 47.
69   Ibid 48.
70   Ibid 5.
71   Ibid 16.
72   Ibid 12.

two concepts are discussed separately 'for the sake of clarity', thus leading to a reproduction of this very dichotomy.

What is the relation between mitigation and adaptation on the one hand, and the issue of poverty on the other? The keyword analysis in part II revealed a conceptualisation of poverty alleviation and adaptation as concurrent policy areas, whereas mitigation is considered to be in a potential conflict with the fight against poverty. On this basis we see poverty alleviation as assigned a place in the resolution or final situation of the adaptation narrative while it is given a place as an additional complication of the mitigation narrative.

This difference transpires clearly in the two following examples, starting with (20), in which the SAG argues for the need to help rural communities adapt to climate change (we only include parts of the discussion, because of its considerable length):

(20) Rural communities with the highest dependence on natural water sources are in KwaZulu-Natal, the Eastern Cape and Limpopo. The former two will probably experience more flooding and water contamination. In addition to these, Limpopo will probably experience more droughts. These are areas with some of **the poorest communities** and under-resourced municipalities with limited capacity and skills to adapt to changing conditions ... In response to these challenges, South Africa will:

   5.7.1 Educate subsistence and small-scale farmers on the potential risks of climate change, and support them to develop adaptation strategies with on-farm demonstration and experimentation. Adaptation strategies will include conservation agriculture practices including water harvesting and crop rotation, and will prioritise indigenous knowledge and local adaptive responses.[73]

Climate change directly influences human settlements, particularly the indigent, requiring adaptive measures to be developed and initiated. Addressing the plight of the poor is thus integral to the reaction component of the adaptation narrative (see Figure 2).

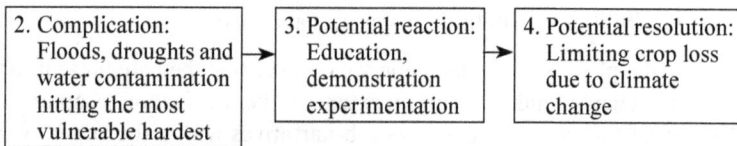

2. Complication: Floods, droughts and water contamination hitting the most vulnerable hardest	→	3. Potential reaction: Education, demonstration experimentation	→	4. Potential resolution: Limiting crop loss due to climate change

Figure 2: Climate change adaptation narrative

In the adaptation narrative, poverty constitutes a problem that is exacerbated by climate change in the complication phase. The reaction, which is composed of policy measures, addresses this problem and leads to a transition in the form of positive effects.

73   Ibid 22.

In order to illustrate how the problem of poverty fits into the narrative of mitigation, let us have another look at example (4), in which the economic impacts of mitigation efforts are discussed:

(4) This policy identifies or sets up processes to identify the optimal combination of actions sufficient to meet the National Climate Change Response Objective. Factors to be considered include not only the mitigation potential, the incremental and direct cost of measures, but also the broader impact on socio-economic development indicators (such as employment and income distribution), our international competitiveness, the cost to **poor households** and any negative consequences for key economic sectors.[74]

The excerpt stresses the need to consider potential negative consequences of mitigation measures. These consequences constitute a secondary complication, leading to the creation of a sub-plot, as illustrated in Figure 3.

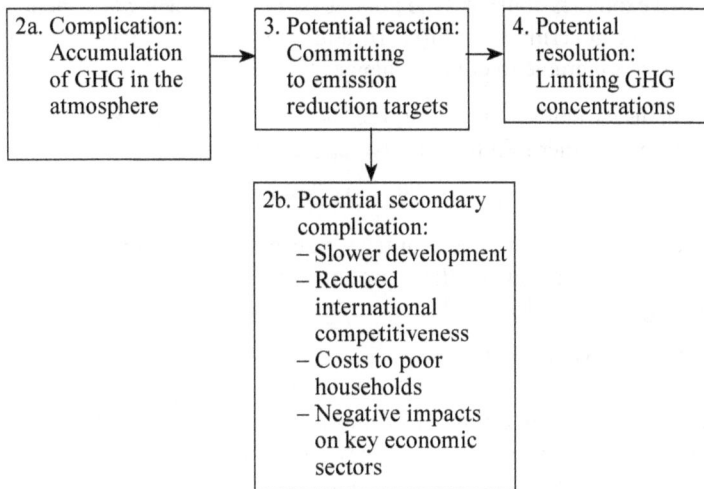

Figure 3: Climate change mitigation narrative

The SAG takes for granted the need to reduce emissions, but the goal of climate change mitigation is compromised by the expected socio-economic consequences of this commitment, including the exacerbation of the plight of the poor, represented by the secondary complication (2b). Through such a conceptualisation of climate change mitigation policies, the White Paper shifts the narrative to the new complication, which needs to be resolved in order for the potential resolution to come about. However, as argued in part II(b), this narrative fails to take into account the fact that the poor already pay a prohibitively high price for electricity from coal powered power plants, because subsidised energy-intensive industry drives up demand.

74   Ibid 27.

Returning to the question of the roles played by different actors, we would maintain that the poor are not given any specific role. They are mentioned, but only as one of many groups which are considered to be particularly vulnerable, as in the paragraph of guiding principles:

(21) Special needs and circumstances – considering the special needs and circumstances of localities and people that are particularly vulnerable to the adverse effects of climate change, including vulnerable groups such as women, and **especially poor** and/or rural women; children, especially infants and child-headed families; the aged; the sick; and the physically challenged.[75]

And in the 'Adaptation' chapter:

(22) … climate-resilient sectoral plans have the potential to directly address the plight of those most impacted by climate change – **the rural poor.**[76]

These two examples construct the poor as passive recipients, or beneficiaries, of benevolent government efforts. In addition the text depicts the poor as particularly vulnerable to climate change:

(23) Climate variability, including the increased frequency and intensity of extreme weather events, will disproportionately affect **the poor.**[77]

This group is also portrayed as incapable agents:

(24) Seventy per cent of the country's **poorest households** live on small-scale farms and few of them produce enough food to feed themselves throughout the year.[78]

In fact there is no portrayal of the poor as active agents, other than generalised assertions on efforts to empower the population. This tendency is reinforced by the frequent use of the word 'vulnerable' to refer to the poor. As with the use of 'poor', there is no real agency, only victims, incapable agents and beneficiaries of mostly government action.

The clearest and most prominent role is attributed to the SAG itself. The SAG is the hero of its own narrative, as demonstrated in the following passages (the first taken from the Introduction and the second from the adaptation chapter):

(25) Against this national context, the South African Government: … Having ratified both the United Nations Framework Convention on climate change (UNFCCC) and its Kyoto Protocol, **will continue to meaningfully engage in the current multilateral negotiations** to further strengthen and enhance the international response to the climate change crisis. The Government specifically aims to continue its efforts to strengthen and ensure the full implementation of the UNFCCC and its Kyoto Protocol through additional multi-lateral rules-based and legally-

75   Ibid 11.
76   Ibid 17.
77   Ibid 9.
78   Ibid 22.

binding international agreements that will come into force after 2012. The Government specifically aims to **continue its efforts to strengthen and ensure the full implementation of the UNFCCC and its Kyoto Protocol** through additional multi-lateral rules-based and legally-binding international agreements that will come into force after 2012.[79]

(26) In response to these challenges, South Africa will:

5.9.1 **Continue to develop and improve** its early warning systems for weather and climate (especially severe weather events) ...

5.9.3 **Continue to promote the development of Risk and Vulnerability Service Centres** at universities, which will, in turn, support resource-constrained municipalities ...

5.9.5 **Maintain**, update and enhance the SARVA as a tool that provinces and municipalities may use to inform their climate change adaptation planning ...

5.9.7 **Develop** mechanisms for the poor to recover after disasters, including micro-insurance.

This *hero* interpretation is corroborated by the polyphonic feature of presupposition, here signalled by the use of the verb 'continue' in both (25) and (26). Let us look at the first example above (our emphasis): '[T]he South African Government: ... **will continue to meaningfully engage in the current multilateral negotiations** to further strengthen and enhance the international response to the climate change crisis.' Through the use of the verb 'continue', the authors presuppose that the government has already started to 'meaningfully engage in the current multilateral negotiations ...'. The effect of such a presupposition is not only to present its content as a given, but also to construct it as accepted by the recipient. In other words, a presupposition represents an underlying content, which is taken for granted, exempted from discussion, thus belonging to a collective voice.[80] In the present example, there is an implicit collective voice, according to which 'the South African Government already engages meaningfully and will continue to do so in the current multilateral negotiations'. The ability to implicitly construct agreement in this way can be rhetorically effective. In the passages above it imposes the status of hero on the SAG.

Provinces and municipalities are also presented as essential to the Climate Change Response Strategy, as heroes in partnership with the SAG. However there are few and only generalised assertions on the mobilisation of finance needed to empower these levels of government. In light of the substantial financial difficulties experienced by municipalities and provinces, the economist Patrick Bond characterises such declarations of local and provincial responsibilities by the SAG as 'buck passing', a practice of delegating responsibilities without providing the means to meet them.[81]

79   Ibid 9.
80   Nølke et al (note 51 above) 33–4.
81   Personal correspondence 26 August 2012.

(c)   The distribution of keywords in the narrative

Going back to the keywords analysed in part II, we can proceed to count their occurrences in relation to the narrative. This poses certain methodological problems. As the notion of *narrative* employed here concerns the conceptual structure of the discourse and not its textual structure, a given stage of the narrative can be realised through single sentences, or even smaller, syntactic elements. This means that a given sentence may encompass several stages of a narrative, as illustrated in the following example:

(27) At the national level, the challenge will be to effectively manage and reduce the economic risks [emerging from the impacts of climate change regulation], and build on and optimise the potential opportunities, and to ensure a smooth and **just** transition to a climate-resilient, **equitable** and internationally competitive lower-carbon economy and society.

This sentence includes a secondary complication, in the form of a presupposition: 'there are economic risks and opportunities resulting from climate regulation'. This stage is accompanied by an outlined reaction (constructed around the verbs 'manage', 'reduce', 'build', and 'optimise'), which is followed by a resolution in the form of an infinitive phrase: 'to ensure a smooth and just transition to a climate-resilient, equitable and internationally competitive lower-carbon economy and society'. The final situation is represented by a noun phrase within this infinitive phrase: 'a climate-resilient, equitable and internationally competitive lower-carbon economy and society'. We see that the secondary complication and the reaction are syntactically intertwined, as are the resolution and the final situation. This means that the identification of a single word as belonging to a given stage of a policy narrative necessitates a thorough linguistic examination, which we will not develop further here.

For the sake of clarity we will limit our examination to the words pertaining to categories (ii) and (iii), which are defined respectively as 'the argumentation of fairness, in which the ethical principle is explicit', and 'the legal argumentation, in which government efforts are presented as not only desirable or necessary but obligatory'. The two keywords in example (27) – 'just' and 'equitable' – belong to category (ii), the argumentation of fairness, and regard the resolution of the narrative, towards a final situation. Let us have a look at two more examples, pertaining to category (ii):

(28) South Africa will build the climate resilience of the country, its economy and its people and manage the transition to a climate-resilient, **equitable** and internationally competitive lower-carbon economy and society in a manner that simultaneously addresses South Africa's over-riding national priorities for sustainable development, job creation, improved public and environmental health, poverty eradication, and social **equality**.

(29) Key outcomes include:

… Outcome 7: Vibrant, **fair** and sustainable rural communities and food security for all.

The different keywords pertaining to category (ii) are used to represent a desirable future or the process leading to it. In fact, as in examples (27), (28) and (29), all occurrences of 'equality', 'equitable', 'equity', 'fair', 'inequalities', 'just' and 'right' are used in assertions regarding future climate change policies, thus pertaining to the reaction, the resolution or the final situation of the narrative. This indicates that the WP describes future actions as guided by social justice or legal rights, but it does not describe the initial situation of socio-economic marginalisation in the same terms. The SAG does recognise the significance of poverty in the initial situation, but this problem is detached from the ethical and legal questions which could be posed to the political party that has been in power for almost 18 years. When 'equality', 'equitable', 'equity', 'fair', 'inequalities', 'just' and 'right' do appear, it is always in the latter phases of the narrative, a future in which the SAG attributes the role of hero to itself.

However, the word 'inequalities' is negatively connoted, in contrast to the others. This has certain consequences, as we can see from the two examples where this word occurs:

(30) Equity – ensuring a fair allocation of effort, cost and benefits in the context of the need to address disproportionate vulnerabilities, responsibilities, capabilities, disparities and **inequalities**.

(31) Spatial planning needs to address historical **inequalities** in land distribution without compromising the ability of the agricultural sector to contribute to food security.

Through the use of 'inequalities', the SAG does formulate existing socio-economic disparities as an ethical problem. However, this problem is not asserted but presupposed. It thus constitutes the background for the assertions themselves, which regard future efforts to combat inequalities. In other terms, these inequalities constitute a backgrounded historical point of departure from which the SAG will act to create a better society. The use of 'inequalities' therefore contributes to the implicit representation of the SAG as a hero in its own narrative.

V   Discussion and Final Remarks

Through the quantitative lexical investigation and the more qualitative narrative analysis, we have pointed at the manner in which the WP constructs the problem of poverty in relation to climate change and climate change response policies. In the lexical analysis of part III we looked at the words 'equality', 'equitable', 'equity', 'fair', 'inequalities', 'just', 'poor', 'poverty', 'right' and 'vulnerable'. We first considered the distribution of these keywords, before subdividing the sentences that contain them into three categories: (i) Poverty alleviation and the protection of vulnerable groups as an argument on its own terms, with no further reference to an ethical or legal principle. This category is defined by the use of 'poor', 'poverty' and 'vulnerable'; (ii) The argumentation of fairness, in which the ethical principle is explicit. This category is defined by the use of 'equality' 'equitable', 'equity', 'fair', 'inequalities' and 'just', but can also include the words belonging to category (i); (iii) The legal argumentation, in which efforts are presented as not only desirable or necessary but obligatory. This category is defined by the use of 'right', but can also include the words belonging to the first two categories.

The analysis led to two main findings. Firstly, in the chapter on 'Adaptation' the frequency of the keywords is six times higher than in the chapter on 'Mitigation'. When the WP does discuss mitigation in relation to poverty alleviation, it presents the two as competing concerns, omitting the current situation of electricity prices being driven up by industrial conglomerates which enjoy energy subsidies. Secondly, we found that a considerable majority (75.5 per cent) of sentences containing one or more of the keywords belong to category (i), and that only one sentence explicitly refers to a policy area in which the authorities have a legal obligation to act in favour of the poor ('the constitutional right to a reliable and safe water supply'). In other words, the SAG formulates the fight against poverty in terms of intentions rather than duties.

The narrative analysis of part IV revealed a complex 'plot', regarding both the components of the narrative and the roles attributed to the different actors. In the complication phase South Africa is portrayed as both victim and villain, as the WP discusses this double role through the use of the concessions 'but', 'however' and 'on the other hand' (see example 15). In a polyphonic perspective, the SAG admits to the country's responsibilities as a 'relatively significant contributor to global climate change', while presenting the fact that 'South Africa is extremely vulnerable and exposed' to the impacts of climate change as being more relevant. As expected from a policy paper, the reaction phase of the narrative is the most prominent. However, the complexities of climate change and the array of policies which are developed to respond to the problem give rise to secondary complications, such as the negative impacts of mitigation measures on poverty alleviation. This is a narrative which fails to take into account the adverse effects of industrial energy consumption on the electricity bills of poor households.

Regarding the role of the poor, the emphasis is on their vulnerability rather than their capacity as active agents in a future climate change response strategy. Their roles vary, as they are cast as both victims, beneficiaries and incapable agents, but they never appear as active agents themselves.

The SAG attributes the clearest role to itself. Both internationally and domestically it constructs itself as the hero of its own narrative. The analysis of a number of examples shows that this role is frequently constructed in the form of presuppositions, which in our polyphonic approach are considered as subtle impositions of general agreement, thus exempting them from discussion.

Less prominent and specific than both the complication and reaction phases are the resolution phase and the final situation, reflecting the significant uncertainties regarding the desired effects of the proposed policies.

Finally, by combining the lexical and narrative analyses we explored the distribution of the different keywords in the different phases of the narrative. The results show a conspicuous absence of the words 'equality', 'equitable', 'equity', 'fair', 'inequalities', 'just' and 'right' in accounts of the initial situation. This leads us to the conclusion that the SAG does not discuss SA's current socio-economic problems in terms of social justice or legal rights. The plight of the poor thus appears to be detached from the ethical and legal dimensions of the SAG's past policies.[82]

In this chapter, we have concentrated on the internal structure and content of the WP, with some contextual factors brought in for justification of our interpretation. However, no text is written in a vacuum. According to the Russian semiotician Mikhail Bakhtin[83] all discourse is fundamentally 'dialogical', because we always try to anticipate and shape reactions to it, and because conversely, all we say is based on previous discourses. No one can avoid reproducing or challenging the concepts and presuppositions of that which has been previously uttered. Many aspects of this encompassing phenomenon known as 'dialogism' concern the linguistically marked voices described by the theory of polyphony, but all words and expressions could be considered as forming a response to anterior discourses. In further research, it would be interesting to investigate more closely which discourses have been the main sources of inspiration for the governmental document analysed here.

Through the investigation of large (electronic) text corpora, it is in fact possible to trace the recurrence of words and collocations.[84] Similarly the National Climate Change Response White Paper could reflect the use of words, expressions and more complex formulations in a larger corpus of national and international climate discourse by political parties, governments, NGOs and international organisations. Such similarities could indicate possible sources having influenced the WP. We are of the opinion that such questions need to be further examined in order to account for the deeper political foundations of texts such as the WP. Corpus-driven approaches may therefore constitute an important component of future research on climate change discourse.

Finally, there are many linguistic issues that could be further examined and developed,[85] thus contributing to the wider knowledge base needed for tackling the challenges of climate change and its legal and ethical dimensions.

---

82   There is one mention of past causes of poverty, but this is attributed to the injustices of the apartheid regime: 'South Africa's cities still reflect apartheid planning with the poorest communities tending to live far away from services and employment' (*White Paper* (note 4 above) 21).

83   M Bakhtin *Speech Genres and Other Late Essays* (1986 [1979]).

84   Collocations are recurrent co-occurrences of words, see J Clear 'From Firth Principles. Computational Tools for the Study of Collocation' in M Baker, G Francis & E Tognini-Bonelli (eds) *Text and Technology. In Honour of John Sinclair* (1993) 271, 277. For an example of collocation analysis, see D Mayaffre & C Poudat 'Quantitative Approaches to Political Discourse: Corpus Linguistics and Text Statistics' in K Fløttum (ed) *Speaking of Europe* (in press).

85   See Fløttum & Dahl (note 6 above); Fløttum (note 9 above); N Koteyko, M Thelwall, B Nerlich 'From Carbon Markets to Carbon Morality: Creative Compounds as Framing Devices in Online Discourses on Climate Change Mitigation' (2010) 32 *Science Communication* 25; B Nerlich, N Koteyko, B Brown 'Theory and Language of Climate Change Communication' (2010) 1 *Wiley Interdisciplinary Reviews: CLIMATE CHANGE* 97.

AN ANALYSIS OF THE HUMAN DEVELOPMENT REPORT 2011: SUSTAINABILITY AND EQUITY: A BETTER FUTURE FOR ALL

Des Gasper, Ana Victoria Portocarrero and Asunción Lera St. Clair

Introduction

A global Human Development Report (HDR) prepared for the United Nations Development Programme (UNDP) appears annually. This series competes with the World Development Report (WDR) series prepared by the World Bank, for the position of the most widely read and influential, agenda-defining, annual report series on international development. The HDR 2000[1] took major steps towards integrating the much longer established framework of human rights with the HDRs' 'human development' conceptual framework, which has been associated especially with Amartya Sen's 'capability approach'.[2] The HDR 2007/2008 presented the problems that anthropogenic climate change poses for development in terms of human rights violations, with reference to power relations at the global level.[3] Runaway climate change 'would represent a systematic violation of the human rights of the world's poor and future generations' by current rich polluters.[4] The HDR 2007/2008 on climate change stood in marked contrast to the equally imposing WDR 2010 on climate change, which adopted a significantly different problem framing and situation diagnosis.[5] The WDR argued that no reduction in economic growth is necessary, and, in line with established World Bank practice,[6] avoided the language of human rights. Desmond McNeill and Asuncion Lera St. Clair have argued that the World Bank's treatment of equity has been primarily

---

*   We would like to thank members of the project on 'Climate change narratives, rights and the poor', and, in particular, Patrick Bond, for useful suggestions on the structure and content of this chapter.

1   UNDP 'HDR 2000: Human Rights and Human Development' (2000).
2   See, for example, A Sen *Development as Freedom* (1999); also see D Gasper 'Human Development' in J Peil & I van Staveren (eds) *Handbook of Economics and Ethics* (2009a) 230, 237; D Gasper 'Human Rights, Human Needs, Human Development, Human Security' in P Hayden (ed) *Ashgate Research Companion to Ethics and International Relations* (2009b) 329, 355.
3   UNDP 'HDR 2007/2008: Fighting Climate Change – Human Solidarity in a Divided World' (2007).
4   Ibid 4.
5   World Bank 'WDR 2010: Development and Climate Change' (2010).
6   D McNeill & AL St. Clair 'The World Bank's Expertise: Observant Participation in WDR 2006 Equity and Development' in D Mosse (ed) *Terms of Reference: the Anthropology of Expert Knowledge and Professionals in International Development* (2011).

instrumental, devoid of an intrinsic value for equity and of one of the central foundations of human rights: the intrinsic equal value of all human beings.[7]

Recent articles have compared these two reports in detail, for they are the major statements on climate change from the two leading international development organisations.[8] In our own analysis,[9] we compared their Overviews using a frame and content-analysis methodology that focused attention on key terms and framing choices. We contextualised this analysis by drawing on research about the institutions' processes of knowledge production and their socio-political environments, and results from earlier work investigating the role of ideas in the multilateral sector. We found something surprising. While the HDR 2007/2008 took a very different stance in comparison to the WDR 2010 in its evaluative language, the policy solutions presented by the two organisations were very largely similar. The HDR did not follow through its human rights-related evaluation into a human rights-based policy perspective. Questions arose regarding the reasons for the inconsistency in the HDR 2007/2008, and how the standpoint of the UNDP – in particular that of the semi-autonomous HDR Office (HDRO) – would evolve.[10]

The HDR 2011 on sustainability and equity provides a relevant opportunity to address these questions. How embedded is a human rights perspective in UN work on human development, and especially in relation to the looming human rights challenge of anthropogenic climate change? The HDR 2011 appeared in the run-up to the UN conference on Sustainable Development, the Rio+20 conference of 2012, and was titled 'Sustainability and Equity: A Better Future for All'.[11] As such it constitutes an attempted major statement, intended to structure the new agenda for sustainability to be adopted by 'the international community'. Responses to climate change challenges must be considered in the context of existing poverty, inequalities and marginalisation. Attempted transitions to sustainable development pathways that do not adequately address climate impacts, the costs and possible negative effects of mitigation strategies, limits to adaptation, or negative synergies among diverse policy options, may lead to negative outcomes for poor communities. In this chapter we ask in which direction did the HDR 2011 move in relation to the schizophrenic stance on climate change seen in the HDR 2007/2008. We also ask to what extent the HDR 2011 met its own demands to integrate equity considerations in discussions about sustainability.

7   McNeill & St. Clair analyse how the processes of producing and negotiating the completion of the WDR 2006 on equity were marked by an equivocal use of the term equity. The WDR placed more emphasis on an instrumental and economistic sense, and downplayed an intrinsic value. See McNeill & St. Clair ibid.

8   K Fløttum & T Dahl 'Different Contexts, Different "Stories"? A Linguistic Comparison of two Development Reports on Climate Change' (2012) 32 *Language and Communication* 14; D Gasper, AV Portocarrero & AL St. Clair 'The Framing of Climate Change and Development: a Comparative Analysis of the Human Development Report 2007/8 and the World Development Report 2010' (2013) 23 *Global Environmental Change* 28, 39.

9   Gasper et al ibid.

10   The HDRO has formal editorial independence in regard to the HDR. However, the office is administratively a part of the UN system and its director is appointed by the UNDP Administrator.

11   UNDP 'HDR 2011: Sustainability and Equity: A Better Future for All' (2011).

Thus this chapter analyses the HDR 2011 with emphasis on the treatment of climate change and the poor, human rights, and North-South relations. We give special attention to the report's relevance to Africa, the poorest continent, and its references to South Africa, a country with a medium Human Development Index (HDI), a growing economy and increasing CO_2 emissions, but high inequality and persistent poverty, in order to illustrate how issues of growth, inequality, climate change and poverty are intertwined and how these are handled in the report. We argue that while the HDR 2007/2008 had taken some important steps in bringing a human rights perspective into the framing of climate change and its importance for the poor, the HDR 2011 on environmental sustainability and equity has stepped back. It downgrades the urgency of climate change impacts and of countering its causes. Moreover, it puts forward a weakened definition of sustainable development compared to the widely used definition of the Brundtland Report,[12] and it distances itself from human rights concerns, from the political barriers to sustainability, and from the structural causes of poverty and inequality as key determinants of social and environmental vulnerability.

Part 2 looks at the HDR 2011's Foreword. Part 3 examines the report's Overview in detail, page by page. Part 4 summarises the framing provided by the report's Overview, its inclusions and exclusions, and compares them with those in the Overviews of the HDR 2007/2008 and the WDR 2010; and then analyses the lexical choices in these Overviews, as a further test of our interpretation of the framings. Part 5 considers the treatment of Africa – the continent likely to be hardest hit by climate change – and particularly of South Africa, in this case in the HDR 2011 as a whole, with reference again to the patterns of inclusion and exclusion of issues and the degree of adequacy of the analysis. Part 6 concludes.

In the remainder of this Introduction we outline the scope of the HDR 2011 and of our methods for investigation. The full report on sustainability and equity is a book-length study of about 125 double-columned pages; the HDR 2011 also contains almost 60 pages of tables of human development data. We concentrate on the report's Overview, because it is a large self-contained document, and is the part of the report that has by far the largest audience and is correspondingly prepared with special attention. Most of the HDR audience only reads the Overview. Given this, we will look at the self-contained version of the HDR 2011 Overview, which is found in the separate Summary file available on the HDR website.[13] It consists of 15 double-spaced pages (in contrast to the 12-page version in the full report). This version incorporates into the Overview selected tables and figures from the full report. The Summary also contains the Foreword, of two pages, and five pages of selected key human development tables. This is unusual – the HDR 2007/2008 Overview did not add such a mass of data – but it matches the line of argumentation in the HDR 2011, much of which is generalised in terms of countries, which are grouped into four categories: very high human development; high human development; medium human development; low

---

12   World Commission on Environment and Development (WCED) 'Our Common Future' (1987).
13   See UNDP 'HDR 2011: Sustainability and Equity: A Better Future for All' (2011) Overview
       <http://hdr.undp.org/en/media/HDR_2011_EN_Overview.pdf>.

human development. The selected tables rank 187 countries in these terms. It is worth adding that the HDR 2011 is a much smaller document than the HDR 2007/2008 on climate change: the earlier study was over 220 pages long, and its Overview was almost double the length of that for 2011, reflecting a more complex and ambitious argumentation.

Table 1: Structure of HDR 2011

(I) The complete report	(II) The Summary
	Foreword by Helen Clark
	Overview
Chapter 1. Why sustainability and equity?	Section 1. Why sustainability and equity?
1. Are there limits to human development?	1. The case for considering sustainability and equity together
2. Sustainability, equity and human development	2. Some key definitions
3. Our focus of inquiry	Section 2. Patterns and trends, progress and prospects
Chapter 2. Patterns and trends in human development, equity and environmental indicators	Section 3. Understanding the links
1. Progress and prospects	1. Environmental threats to selected aspects of human development
2. Threats to sustaining progress	• Bad environments and health – overlapping deprivations
3. Success in promoting sustainable and equitable human development	• Impeding education advances for disadvantaged children, especially girls
Chapter 3. Tracing the effects – understanding the relations	• Other repercussions
1. A poverty lens	2. Disequalizing effects of extreme weather events
2. Environmental threats to people's well-being	3. Empowerment – reproductive choice and political imbalances
3. Disequalizing effects of extreme events	• Gender inequality
4. Disempowerment and environmental degradation	• Power disparities
Chapter 4. Positive synergies – winning strategies for the environment, equity and human development	Section 4. Positive synergies – winning strategies for the environment, equity and human development
1. Scaling up to address environmental deprivations and build resilience	1. Access to modern energy
2. Averting degradation	2. Averting environmental degradation
3. Addressing climate change – risks and realities	
Chapter 5. Rising to the policy challenges	Section 5. Rethinking our development model – levers for change
1. Business-as-usual is neither equitable nor sustainable	1. Integrating equity concerns into green economy policies
	2. A clean and safe environment - a right, not a privilege
	3. Participation and accountability
	4. Financing investments: where do we stand?
	5. Closing the funding gap: currency transaction tax…

(I) The complete report	(II) The Summary
2. Rethinking our development model- levers for change 3. Financing investment and the reform agenda 4. Innovations at the global level Tables	6. Reforms for greater equity and voice Tables

We will present a frame-analysis of what has been included and what excluded in the HDR 2011 Overview, and use lexical analysis (study of word choice) to examine how what is included has been characterised and described.[14] As an initial example, Table 1 presents a comparison of the scope of the full report and of its Overview. Naturally there is a close correspondence, but missing from the Overview are sections corresponding to those in the full report on 'Are there limits to human development?', 'A poverty lens', 'Scaling up to address environmental deprivations and build resilience', 'Addressing climate change – risks and realities', and 'Business-as-usual is neither equitable nor sustainable'. In the other direction, 'Access to energy' appears as a section in the Overview but not in the full report. Table 1 highlights the discrepant sections. The Overview is the public face of the report, the section read by most journalists, administrators or students. Its set of topics seems blander, more technocratic, and less challenging to business-as-usual. The exclusion of the indicated sections is relevant, for some of them contain information that helps in nuancing and even countering some of the assertions made in the Overview, as we will explain later. The exclusions are also relevant for the way in which the challenges posed by climate change in Africa are depicted, favouring a focus on hoped-for win-win-win solutions rather than addressing the difficult structural causes of people's vulnerability.

Beneath the level of section headings, what do the sections contain? Whereas the full report includes a modest but substantive section on human rights to a safe and sufficient environment,[15] using Martha Nussbaum's language of people's underlying 'rights to bodily health and integrity and to enjoyment of the natural world',[16] the Overview races through 'A clean and safe environment – a right not a privilege' in three paragraphs. The condensation ratio is far greater here than, for example, for the technical economic topic of the Multidimensional Poverty Index (MPI), which receives nine paragraphs in the Overview.[17] The treatment of rights in the Overview shrinks to a mention

---

14   For outlines of the methods employed, see RJ Alexander *Framing Discourse on the Environment:*
     a Critical Discourse Approach (2009); R Schmidt 'Value-Critical Policy Analysis' in D Yanow
     & P Schwartz-Shea (eds) *Interpretation and Method* (2006); T van Dijk (ed) *Discourse Studies. A*
     Multidisciplinary Introduction (2011); D Yanow *Conducting Interpretive Policy Analysis* (2000).
15   HDR 2011 (note 11 above) 86–8.
16   Ibid 86.
17   See UNDP 'HDR 2011: Sustainability and Equity: A Better Future for All' (2011) Summary
     <http://hdr.undp.org/en/media/HDR_2011_EN_Summary.pdf> 7–8.

of how legalised rights 'can be effective',[18] without attention to their grounding and justification.

Overall, our chapter aims to throw light on: what principles guide the work on sustainability by the official champion of the human development approach, the HDRO in the UN, and the degree to which human rights concerns and the interests of poor people are as yet reflected there; whether the incomplete steps towards incorporating such concerns that were taken in the HDR 2007/2008, with respect to threats arising from climate change, have been maintained and taken further, or lost and replaced by something else; and, if not maintained and extended, what are the implications of such a failure to adopt a human rights orientation in response to the threats brought by climate change, for the human rights of poor people everywhere, and not least in South Africa.[19]

II FOREWORD TO THE HDR 2011

The Foreword is signed by Helen Clark, former prime minister of New Zealand, Administrator of the UNDP since 2009 and the first woman to lead the organisation. It opens by pointing to the importance of seeking consensus at the 2012 Rio de Janeiro global summit conference, on 'global actions to safeguard the future of the planet and the right of future generations everywhere to live healthy and fulfilling lives'.[20] It states that '[T]his is the great development challenge of the 21st century'. The emphasis here is 'on the planet' and on 'future generations', rather than on current generations of people suffering the consequences of climate change and poverty. Intragenerational justice is considered in the course of the report, but it is not part of the 'great development challenge' posed in this introductory paragraph.

The Foreword presents the report's main idea as follows: 'sustainability is inextricably linked to basic questions of equity – that is, of fairness and social justice and of greater access to a better quality of life'.[21] And it continues by stating that sustainability 'is fundamentally about how we choose to live our lives, with an awareness that everything we do has consequences for the 7 billion of us here today, as well as for the billions more who will follow'.[22] The 'we' here is all humanity. This initial treatment of humanity as one is important, though it needs to be followed by disaggregation, otherwise it can erase issues of power imbalances between people in the global North, whose actions have had bigger consequences in the lives of everybody else in the world, and people of the global South, whose decisions are less influential in the lives of those in powerful positions.

The word 'power' is indeed mentioned in the fifth paragraph of the Foreword, referring to two levels at which power disparities shape the

18 Ibid 12.
19 See also D Gasper 'Climate Change – the Need for a Human Rights Agenda within a Framework of Shared Human Security' (2012) 79 *Social Research: an International Quarterly of the Social Sciences* 983, 1014.
20 HDR 2011 (note 11 above) ii.
21 Ibid.
22 Ibid.

burdens and deprivations of poor people. The first level is the nation, where 'power disparities and gender inequalities ... amplify the effects of income disparities'; and the second level is global, where 'governance arrangements often weaken the voices of developing countries and exclude marginalized groups'.[23] Although power imbalances are identified at these two levels, only one of them is then mentioned in the action alternatives proposed by UNDP's Administrator in the Foreword: the national. The Foreword specifically suggests that:

> [s]uccessful approaches rely on community management, inclusive institutions that pay particular attention to disadvantaged groups, and cross-cutting approaches that coordinate budget and mechanisms across government agencies and development partners.[24]

The emphasis is on the national and sub-national, and the role of the global is mainly for financing, which has been its traditional official role. The Foreword does not follow up the impact of power imbalances at the global level; instead it goes further in relation to the national level, advocating investments in renewable energy, water and sanitation and reproductive health care. The Foreword closes with messages about the opportunity that Rio 2012 represented to reach a 'shared understanding' of how to move forward.[25] A key phrase, which captures the role of the global North in this 'shared understanding' is: 'Hope rests on new climate finance': more funds, and directing them 'towards the critical challenges of unsustainability and inequity'.[26]

According to the Foreword, '[u]nderstanding the links between environmental sustainability and equity is critical if we are to expand human freedoms for current and future generations'.[27] The emphasis is on 'expanding' human freedoms, not explicitly on constraining the excessive freedoms of some groups that bring damage to fundamental freedoms of others. The liberal conception that development is about how individuals 'choose to live [their] lives', which we cited earlier, can sometimes be in tension with sustainability and equity and with seeking common and public goods. In general terms, equity and sustainability require, on the one hand, actions to expand some people's freedoms from fear and from want (meaning non-fulfilment of basic needs), and on the other hand actions to restrict people's freedoms to accumulate at the expense of damaging others and the environment. We will see that the emphasis of the report is on the first set of actions, focusing on the poor and disadvantaged in the world, and not on the actions of the rich and on the need to design also some restrictions to cope in particular with the emissions problem. The Overview declares that 'people's chances at better lives should not be constrained by factors outside their control',[28] a sentiment that is intended to apply to the poor; but that is in danger of being converted

---

23   Ibid.
24   Ibid.
25   Ibid iii.
26   Ibid
27   Ibid ii.
28   Ibid 1.

into a slogan that provides protection for the rich against taxation and against regulation to control their actions that harm others.

Sometimes a report's Foreword pulls in a different direction to the report itself, when an organisational apex dissociates itself from the ideas of a subsidiary. Some such tensions existed between the human rights perspective in the HDR 2007/2008 and the Foreword written by the then UNDP Administrator, long-time World Bank official Kemal Dervis. But, whereas the HDR 2007/2008's title was 'Fighting Climate Change: Human Solidarity in a Divided World', the title of the HDR 2011 and of its Overview is far blander: 'Sustainability and Equity: A Better Future for All'. Reference to human solidarity in order to counter global divisions, and to the language of human rights, which was often used in the HDR 2007/2008, are very largely replaced by the vaguer language of 'equity', as used by the WDR 2010.[29] Similarly, whereas the cover of the HDR 2007/2008 portrayed an image of the world in red, with a silhouette of a person looking at the world, the HDR 2011 cover has a geometrical figure: a square that balances on one vertex. It has no personal element and appeals not to compassion and responsibility but to a purely abstract notion of equity. The cover matches an equally abstract diagram used in the report to analyse possible trade-offs between equity, sustainability and expansion of valued freedoms.

From reading the Foreword we have identified some of the issues whose treatment in the HDR 2011's Overview deserve examination: the degree of attention to regulating excessive and damaging freedoms, in addition to promoting expansion of freedoms; the degree of attention to the activities and responsibilities of rich persons, not only poor persons; and the degree of attention to the development, activities and responsibilities of rich countries, not only poor countries.

III MAIN THEMES IN THE HDR 2011 OVERVIEW

The Overview opens by emphasising the links between sustainability, equity and human development, stating that environmental degradation intensifies inequality through adverse impacts on already disadvantaged people, and that inequalities in human development amplify environmental degradation. The report aims to point to pathways that can achieve human development together with sustainability and equity.[30]

(a) Section on 'Why sustainability and equity?'

The opening section proposes that environmental sustainability and equity 'are fundamentally similar in their concern for distributive justice'[31] yet even today these topics are approached separately in many cases. Sustainability

29 Gasper et al (note 8 above).
30 Overview (note 13 above) 2. The HDRs of the 1990s that elaborated on the concept of human development incorporated sustainability and equity, as in the HDRs of 1996 and 1997. The usage in the HDR 2011 appears often to shrink the concept to, for example, what is covered by the Human Development Index.
31 Ibid 1.

reflects a concern for future generations, and equity reflects a concern for those in current generations who are disadvantaged. Next the concept of human development is presented, using Sen's conceptualisation, ie expansion of the field of attainable valued outcomes. As noted earlier, the focus on expanding some people's choices leaves out of focus the question of constraining some others' freedoms. This is of central importance as sustainability refers to ecological limits that logically imply limits on humans' appropriate freedoms to use nature. Not surprisingly, the formulation then given by the report of the concept of sustainable human development – 'the expansion of the substantive freedoms of people today while making *reasonable efforts* to avoid *seriously* compromising those of future generations',[32] represents a major weakening of the 1987 Bruntland Commission definition. That referred to development, which 'meets the needs of the present without compromising the ability of future generations to meet their own needs';[33] it had a focus on outcomes not just efforts and left no room for equivocation about what is 'serious' damage.

The report advocates an approach to environmental sustainability that 'favours the position of preserving basic natural assets and the associated flow of ecological services',[34] and argues that this aligns with human rights-based approaches to development. But the argument is a mere claim, since no specifics are given to demonstrate how this focus on preserving environmental assets has direct or indirect consequences for rights. At the same time, it makes the case for a framework that acknowledges that sustainability and equity are not necessarily mutually reinforcing, 'for example if [pro-environment measures] constrain economic growth in developing countries'.[35] The report encourages us to give special attention to trade-offs between these objectives and to identifying policies that show 'positive synergies': what they call 'win-win-win solutions that favour sustainability, equity and human development'.[36] 'Win-win' is a concept for describing when more than one group gains, but here it is adapted to describe how more than one objective can be promoted. It is a part of managerial jargon widely used in dominant framings of climate and development, such as that presented in the WDR 2010. Moreover, there is a switch in the definition of sustainable development from referring to the lives of specific groups of people to referring to national averages and aggregates for achievement of objectives. This stands in contrast to the humanistic language chosen in the HDR 2007/2008, which repeatedly openly emphasised the lives and livelihoods of recognisable groups of persons, including in its Overview: 'rural communities in Bangladesh, farmers in Ethiopia and slum dwellers in Haiti'.[37] Such language has disappeared in the HDR 2011 Overview.

32   Ibid 2 (emphasis added).
33   WCED (note 12 above).
34   Overview (note 13 above) 2.
35   Ibid.
36   Ibid.
37   Gasper et al (note 8 above).

(b)   Section on 'Patterns and trends, progress and prospects'

This section in the Overview argues that: the links between the HDI (particularly its income component) and environmental degradation are not linear; the poor are affected disproportionately by climate change; but win-win-win solutions can be enacted, so growth does not have to be unequal and unsustainable.

It first examines a range of predictions related to environmental degradation and considers the implications for human development. According to two simulations prepared for the report of the consequences if climate change is not properly addressed, the global HDI in 2050 will be eight or 15 per cent lower than in a baseline scenario which avoids those consequences, due to negative impacts on agricultural production, access to clean water and improved sanitation, etc. 'These projections suggest that in many cases the most disadvantaged people bear and will continue to bear the repercussions of environmental deterioration, even if they contribute little to the problem.'[38] The scale and significance of these repercussions is in fact seriously concealed by an aggregated index like the HDI, which sums up achievements for all inhabitants of a country or region. Such an index is misleading and inadequate for conveying the scale of impacts on poor people, many of whom may fare far worse than does the national or regional average. Unlike the HDR 2007/2008 which spoke openly of the tens and hundreds of millions of poor people who would be seriously and sometimes fatally hit by these 'repercussions', the HDR 2011 loses such information in the almost useless hyper-aggregated category of global HDI. The sacrifice of many people at the margins of global society becomes virtually invisible under the veil of the global HDI figure.

Moreover, there is no balanced treatment of the structural causes of current non-climatic and non-environmental stressors that lead towards low human development levels in the future. This is in striking contrast with current scholarship on the human dimensions of climate change and on adaptation research, for example. It is widely recognised that structural factors and drivers of poverty are central hindrances for future resilience, limit adaptation options and may even lead to maladaptation.[39] A discussion on specific constraints towards sustainability posed by existing vulnerabilities is absent in chapters of the report where these constraints would logically play a central role. There is one single reference to poverty in the chapter 1 'Why Sustainability and Equity?' and zero references to Africa, the poorest continent. Yet this is the chapter that discusses the limits to human development. The main references to Africa and to poverty are in relation to what the report calls win-win-win strategies. This centres the reader's attention on changes in the South towards sustainability rather than on required changes in the unsustainable rich North,

38   Overview (note 13 above) 3.
39   J Barnett & S O'Neill 'Maladaptation' (2010) 20 *Global Environmental Change* 211; L Jones & E Boyd 'Exploring Social Barriers to Adaptation: Insights from Western Nepal' (2011) 21 *Global Environmental Change* 1262; E Marino & J Ribot 'Adding Insult to Injury: Climate Change and the Inequities of Climate Intervention' (2012) 22 *Global Environmental Change* 323.

and it avoids contextualising the limits to such a transition that are posed by low human development.

Another notable choice in aggregation occurs when the Overview explains that 'three quarters of the growth in emissions since 1970 comes from low, medium and high HDI countries', adding though that 'overall levels of greenhouse gases remain much greater in very high [HDI] countries'.[40] It is not clear why the report puts high HDI countries in the same group with medium and low HDI countries; separating these countries could be more revealing. For example, what is the share of low HDI countries in the growth of emissions since 1970? Implicitly the report responds instead to a discussion agenda set by the remaining group, the very high HDI countries. At the same time it makes clear that '[e]missions per capita are much greater in very high HDI countries than in low, medium, and high HDI countries combined, because of more energy-intense activities'.[41] A person in a very high HDI country accounts for more than 30 times the carbon dioxide emissions per person in a very low HDI country. In addition, the Overview notes that the current economic relationships between countries allow the allocation of carbon-intensive production to poor countries while its output is exported for consumption in rich countries.[42] Deepening the comparisons of the environmental impacts attributable to different countries in order to reflect this could easily have been done, but was not.

The Overview suggests next that 'where the link between the environment and quality of life is direct, as with pollution, environmental achievements are often greater in developed countries, [and] where the links are more diffuse, performance is much weaker'.[43] It gives three findings that support this general view. First, household environmental deprivations (indoor air pollution, inadequate access to clean water and improved sanitation) decline as the HDI rises. However, although these household deprivations can be seen as environmental issues, they should not be compared with urban air pollution or greenhouse gas (GHG) emissions. They are poverty issues, whereas urban air pollution and GHG emissions are consequences of development-as-usual models. Household environmental deprivations are issues that should be addressed as poverty-related or developmental problems, while the main environmental issues at stake in the sustainability debates are the ones in which developed countries are more involved and developing countries suffer the first and more intense impacts of climatic change.

The full report's section on 'a poverty lens' – which as we saw lacks a counterpart section in the Overview – discusses the variables used as indicators of environmental deprivations: water, sanitation, cooking fuel. The section admits that:

> [t]he three environmental deprivations were selected as the best *comparable* measures across countries, but other environmental threats and direct impacts of climate change may be equally or

40   Overview (note 13 above) 3.
41   Ibid.
42   Ibid.
43   Ibid 4.

more acute at the local or national level. Flooding may be a more pressing concern for poor households in Bangladesh, for example, than access to water.[44]

By selecting three variables that are more linked to poverty than to impacts of unsustainable behaviour on the poor, the focus shifts to developing countries and away from the sources of unsustainability. If floods, for example, were the focus we would consider more the causal links between GHG emissions and environmental and social disasters. The examples used, such as access to sanitation and modern cooking fuels, do not make those links visible, and hide from view the consumption and production patterns in rich countries.

Second, environmental risks that have community effects (like urban air pollution) seem to rise and then fall with development. However, one can note that the inverted U-shape partly arises because some of the activities that create urban air pollution are moved from richer to poorer countries. Environmental risks that have global effects (notably GHG emissions) typically rise with the HDI. According to the report, '[t]he HDI itself is not the true driver of these transitions', and while incomes and economic growth 'have an important explanatory role for emissions ... the relationship is not deterministic either', since for example 'large-scale commercial use of natural resources has different impacts than subsistence exploitation'.[45] 'Several countries have achieved significant progress both in the HDI and in equity and environmental sustainability',[46] the win-win-win solution.

The report shows a table of good performers, which includes Costa Rica, Germany, Philippines and Sweden.[47] It concludes that 'across regions, development stages and structural characteristics countries can enact policies conducive to environmental sustainability, equity and the key facets of human development captured in the HDI'.[48] The win-win-win solutions that these countries illustrate are in terms of three criteria: global threats, local impacts, and equity and human development. The global threats considered are GHG emissions, deforestation and water use. Sweden and Germany comply with all the criteria except for GHG emissions. That the report chooses Germany and Sweden as examples of win-win-win policies, even though neither performs adequately by the GHG emissions criterion,[49] shows its downgrading of climate change and of the global poor. Thus not included in its estimation of win-win-win performance is attention to the poorer groups in tropical and subtropical zones who are those most affected by climate change impacts of emissions in Germany, Sweden and elsewhere in the North – as was eloquently described in the HDR 2007/2008.

Immediately after presenting Germany and Sweden as good examples, the Overview exposes 'environmental deterioration on several fronts [elsewhere], with adverse repercussions on human development',[50] such as land degradation

44   Ibid 49 (emphasis added).
45   Ibid 4.
46   Ibid.
47   Ibid 5.
48   Ibid.
49   See Table 1 of the Summary (note 17 above) 5.
50   Overview (note 13 above) 5.

due to soil erosion and overgrazing, unsustainable water use in agriculture, and deforestation, described as a 'major challenge', especially in Latin America and the Caribbean, sub-Saharan Africa, and the Arab States.[51] This geographical focus is defended in terms of the bigger impacts of environmental degradation and climate change on those areas,[52] given the higher dependency of poor people on natural resources to make a living, the higher dependency of women on forests and fishery, the greater adverse consequences of environmental degradation faced by women in poor countries because they are disproportionately involved in subsistence farming and water collection, and the greater adverse impacts on indigenous people who also rely heavily on natural resources. The Overview notes in passing that '[e]vidence suggests that traditional practices can protect natural resources, yet such knowledge is often overlooked or downplayed'.[53] This major issue itself then receives no further attention. Even in the full report there are only two brief references (in chapter 4). Also unexplored is the evidence that high health achievements may be correlated with lower carbon emissions whereas high incomes are not.[54]

While the Overview's global analysis allows us to see the role of high HDI countries in climate change (specifically their responsibility for GHG emissions), the local analysis that follows, and the emphasis on problems of the use of land, water and forests, allows the report to shift our attention to low-income regions and away from regions such as the United States and Europe, where the biggest problems of GHG emissions have been and continue to be. The report focuses on national and local levels in less developed countries, downgrading the other side of the coin: the need for changes in highly developed countries.

(c)   Section on 'Understanding the links'

This section of the Overview considers links between environment and equity, including with reference to gender roles and empowerment. It starts by emphasising that most disadvantaged people carry a double burden of deprivation: they are more vulnerable to the wider effects of environmental degradation, and they also have to cope with 'threats to their immediate environment posed by indoor air pollution, dirty water and unimproved sanitation'.[55] But the report focuses more on diverse poverty issues than on the enormous longer-run implications of climate change. 'Climate change' is referred to only 23 times in the Summary for the HDR 2011, compared to 175 times in the Summary for the HDR 2007/2008.[56] The latter is somewhat longer (31 pages compared to 26 pages), and the former covers more aspects of environment; but still a major shift in focus is evident, which is seen in

---

51   Ibid 5.
52   Ibid 6.
53   Ibid.
54   JK Steinberger, J Timmons, R Peters & G Baiocchi 'Pathways of Human Development and Carbon Emissions Embodied in Trade' (2012) 2 *Nature Climate Change* 81.
55   Overview (note 13 above) 7.
56   We include in both cases the report's Forewords.

the frequency of usages of the term, 12.6 uses per 1,000 words in the HDR 2007/2008 versus 2.8 in the HDR 2011. This trend is maintained in the full report. While climate change is used 1,247 times in the HDR 2007/2008, it is mentioned 18 per cent as often in the HDR 2011 (229 times). Furthermore, most of the central points of an important section of chapter 2 that addresses climate change are missing in the Overview. That section makes bold statements regarding consumption patterns and carbon emissions, carbon-intensive production, trade and shifts of carbon emissions, and who the net importers are of carbon and wood.

In contrast, the section in the Overview elaborates on 'the pervasiveness of environmental deprivations among the multi-dimensionally poor'.[57] Indeed it argues that these deprivations – in access to modern cooking fuel, clean water, and basic sanitation – 'disproportionately contribute to multidimensional poverty'.[58] But what deserves to be called an environmental deprivation and what is simply an income poverty or economic poverty issue? No doubt, if these economic poverty issues are also called environmental deprivations, it is logical to say that the environmental deprivations contribute to poverty. But to focus on these 'environmental deprivations' at the expense of looking at impacts of climate change appears misguided.

The Overview then describes a series of familiar links, from the 'environmental deprivations' and other environmental degradation, to deterioration in poor people's capabilities and lives. Less routinely, it notes that extreme weather events have disequalizing effects, and greater exposure to extreme weather events substantially reduces a country's HDI.

Also novel, but problematic, is a proposition that '[t]ransformations in gender roles and empowerment have enabled some countries and groups to improve environmental sustainability and equity, advancing human development'.[59] In particular, 'in countries where effective control of reproduction is universal, women have fewer children, with attendant gains for maternal and child health and reduced greenhouse emissions' [and] 'evidence suggests that if all women could exercise reproductive choice, population growth would slow enough to bring greenhouse gas emissions below current levels'.[60] Yet, as noted earlier in the Overview, emissions per capita are very unequal: CO_2 emissions per person in a very high HDI country are 30 times more than for a person in a very low HDI country.[61] This means that, even if it was true that overall emissions would decrease if women in poor countries had less children, the impact of such policies would be far less than from the reduction of emissions in rich countries. Considering this, it is perhaps bizarre to suggest population control in the global South in order to reduce GHG emissions. The data lead us rather towards a call for changes in the patterns of production and consumption in the global North. Further, the data show that precisely in the

57   Overview (note 13 above) 7.
58   Ibid.
59   Ibid 9.
60   Ibid 10.
61   Ibid 3.

global North, where more gender equality has been achieved, dramatically higher per capita GHG emissions are produced.

This has implications too for two further links proposed in the section: first, that women's political participation has important implications for sustainability and equity:

> because women often show more concern for the environment, support proenvironmental policies, and vote for proenvironmental leaders, their greater involvement in politics and in nongovernmental organizations could result in environmental gains, with multiplier effects across all the Millennium Development Goals.[62]

Given the patterns observable in rich countries with relatively high participation by women but also very high GHG emissions, the proposed link seems at best a weak one. Similarly, the report states that:

> political empowerment at the national and subnational levels has been shown to improve environmental sustainability [and that] while context is important, studies show that democracies are often typically more accountable to voters and more likely to support civil liberties.[63]

Yet many countries with stronger democracies are those with stronger CO_2 footprints, which refutes any simple relationship between democracy, sustainability and equity. The section itself adds that even within democratic countries the needs and views of the most affected groups are often not reflected in policy priorities, and that the most important issue to solve is power imbalances.

Regarding what the section does not cover, issues about imbalances between countries in terms of power to set the agenda on climate change are not mentioned. The Foreword's warning that power disparities at the global level 'weaken the voices of developing countries and exclude marginalized groups',[64] is not followed up.

(d) Section on 'Positive synergies – winning strategies for the environment, equity and human development'

This section of the Overview discusses the win-win-win strategies that are to be created by governments, civil society, private sector actors and development partners. Although effective solutions should be context specific, there are some principles that could work across countries, according to the Overview. At the local level, it stresses the need for inclusive institutions, and at the national level it aims for scaling-up of successful innovations and policy reform. The international/global level remains absent. The emphasis is on what has to be done at local and national levels, and specifically in developing countries. In the full report too, while some additional references are made in relation to power imbalances in decision-making processes at the global level, when it comes to the solutions these imbalances are only addressed by reference to new financing mechanisms.

---

62   Ibid 10.
63   Ibid.
64   Summary (note 17 above) ii.

One specific example of proposed win-win-win strategies concerns 'Access to modern energy'. The subsection starts by asking: 'Is there a trade-off between expanding energy provision and carbon emissions?', and the response is 'not necessarily', since '[t]here are many promising prospects for expanding energy without a heavy environmental toll'.[65] For example, off-grid decentralised options are, according to the report, 'technically feasible for delivering energy services to poor households and can be financed and delivered with minimal impact on the climate'.[66] 'Providing basic modern energy services provision for all would increase carbon dioxide emissions by only an estimated 0.8 percent'.[67] The challenge is to expand access to energy supply with renewables at a scale and speed that will improve poor people's lives. But while the report addresses the problem of avoiding increased impact on the climate, it does not address the problem of how to decrease that impact from current unsustainable levels. Neither does it address questions of unequal *access* to *existing* energy sources, as we shall see in its discussion of South Africa.

In a section titled 'Scaling up to address environmental deprivations and build resilience' (which has no counterpart in the Overview), the full report takes a conservative position in relation to cuts of emissions, stating that '[p]olicies to cut emissions nationally entail both potential advantages and concerns about equity and capacity'.[68] Table 4.1 lists policy instruments to cut carbon dioxide emissions: cap-and-trade permits, emissions targets, taxes or charges, subsidies for renewables, subsidy cuts, performance standards, technology standards, and better information. The only ones with positive comments on 'equity aspects' are: subsidy cuts and better information. However, the report does not say where these cuts are going to happen, whether in developing or developed countries, nor is there any treatment of intra-country emissions inequalities across diverse social groups and sectors. The report's focus on action in developing countries, where cutting of emissions may sometimes have an inverse relationship with equity (due to the relation that the report has shown between development and access to water, sanitation and energy), leads to claims which are then excessively generalised, and implicitly used to reduce the pressure for adjustments in advanced economies, who are anyway not openly discussed.

More broadly, the proposed win-win-win measures for averting environmental degradation include expanding reproductive choice, promoting community forest management, and adaptive disaster response, among others. Reproductive choice is presented as a precondition of women's empowerment, and, for reasons touched on earlier, as favourable for averting environmental degradation. Community forest management could, proposes the report, redress local environmental degradation and mitigate carbon emissions, though one must beware exclusionary processes inside communities through

65   Overview (note 13 above) 11.
66   Ibid.
67   Ibid.
68   Ibid 69.

which disadvantaged groups may be marginalised. The third measure mentioned is the development of equitable and adaptive disaster responses and innovative social protection schemes.

The options mentioned in the report are important, but remain diffuse and indirect. They do not directly address issues of climate change. Moreover they do not provide long-term vision and fail to take into account the possibility of a high-end increase in average temperatures, which will make most of the win-win-win proposals in this section unfeasible. In addition, these are again measures applied at the very local levels in developing countries, excluding the global level, as if measures in developed countries had no impact on the environment, although the data given earlier has made clear that such measures would have a stronger impact in comparison with measures in developing countries. Thinking about measures in developing countries is essential, but leaving invisible the responsibilities of developed countries on these issues is unjustifiable.

(e)   Section on 'Rethinking our development model – levers for change'

This section in the Overview presents a 'new vision for promoting human development through the joint lens of sustainability and equity'.[69] For the local and national levels, the report stresses 'the need to bring equity to the forefront of policy and programme design and to exploit the potential multiplier effects of greater empowerment in legal and political arenas'.[70] For the global level, the report highlights 'the need to devote more resources to pressing environmental threats and to boost the equity and representation of disadvantaged countries and groups in accessing finance'.[71]

Although this section declares it is about rethinking our development model, these proposals are from a very longstanding development agenda. The global and international levels are again presented only as spaces for mobilisation of finances and not as spaces where rethought structures and systems can be designed and implemented. This section (as well as the corresponding full chapter in the report) presents the biggest challenges for a sustainable future as concerning changes in the global South, not changes in models of development in the North.

More specifically, the report presents the following ideas:[72] (1) Integrating equity concerns into green economy policies: with reference not only to people's incomes but to non-income dimensions of well-being, indirect effects of policy, compensation mechanisms for adversely affected people, and the risks of extreme weather events; (2) Embedding environmental rights in national constitutions and legislation, as well as empowering institutions to work with these rights; (3) Participation and accountability: 'greater empowerment can bring about positive environmental outcomes equitably',[73]

---

69   Ibid 12.
70   Ibid.
71   Ibid.
72   Ibid.
73   Ibid.

by making national institutions accountable; for example, it stresses the need to 'strengthen the possibilities for some traditionally excluded groups, such as indigenous peoples [and women], to play a more active role'.[74]

These ideas bring together what has been already presented in the Overview, rather than transcend it as the section's title could have implied. The attractions and limitations remain the same. First, the links drawn with the environment and especially with climate change are vague. Second, the report takes for granted the concept of green economies as a positive one without distributional consequences (in contrast, see the United Nations Research Institute for Social Development (UNRISD) study on the social dimensions of the green economy).[75] Third, these measures focus on local and national levels in developing countries, and do not address the main problems around climate change, which is probably the main threat to sustainability. An exception to this limitation is found in the following subsection, which proposes a currency transactions tax.

(i)   Subsection on 'Closing the funding gap'

This subsection argues that the investments needed are large but not disproportionate in relation to current spending on other sectors, let alone on the military and least of all in comparison with speculative money flows. In order to reduce the funding gap, a currency transaction tax is suggested. According to the report, 'at a very minimal rate (0.005 percent), and without any additional administrative cost, the currency transaction tax could yield additional annual revenues of about $40 billion'.[76] An additional mechanism suggested is the monetization of part of the IMF surplus Special Drawing Rights.

In this one area the report makes a proposal, the currency transaction tax, that matches the global scale and character of the environmental dangers highlighted by repeated international studies like the Millennium Ecosystem Assessment (2005), the UNEP Global Environmental Outlook studies (2007, 2011) and the reports of the IPCC (2007).[77] In other areas it remains significantly limited in its policy agenda. For example, when it argues 'for large transfers of resources to poor countries, both to achieve more equitable access to water and energy and to pay for adapting to climate change and mitigating its effects',[78] the term 'mitigation' has, without comment, evolved from its normal use in climate discussions, namely the reduction of emissions of GHGs, to instead refer to mitigation of the effects of the resultant climate change. Remarkably, the Overview addresses only the effects of climate change, not the causes, i.e. matters that especially concern the behaviour and

74   Ibid.
75   UNRISD 'Social Dimensions of Green Economy and Sustainable Development' (2012).
76   Overview (note 13 above) 13.
77   Millennium Ecosystem Assessment (2005) <http://www.unep.org/maweb/en/index.aspx>; UNEP Global Environmental Outlook studies (2007, 2011) <http://www.unep.org/geo/>; and the reports of the IPCC (2007) <http://www.ipcc.ch/publications_and_data/publications_and_data_reports.shtml#1>.
78   Overview (note 13 above) 13.

responsibilities of governments in wealthy countries and of affluent consumers and corporations in all countries.

The Overview uses the terms 'mitigate', 'mitigation' or 'mitigating' only five times, and never specifically to refer to reduction of GHG emissions in rich countries or by rich producers and consumers. Indeed only one of the five uses even permits such reference; the others are about developing countries or 'mitigation of effects'. These terms are used 27 times in total in the full report, plus 10 times in the References and Notes. From the 27 uses, 10 are clearly related to mitigation of the effects of climate change, or are used ambiguously, and 11 more are linked to financing mitigation or mitigation costs. Four cases refer to mitigation of carbon emissions, but it is not clear in which countries, and two uses are not related to climate change at all. Considering that the report mostly focuses on developing countries, it can be inferred that usage of the term is largely linked to mitigation of effects of climate change or mitigation of carbon emissions taking place in developing countries.

(ii)   Subsection on 'Reforms for greater equity and voice'

The last subsection of the Overview discusses reforms for greater equity and voice. It remarks that accountability is not sufficient but is a necessary condition for 'building a socially and environmentally effective global governance system that delivers for people'.[79] Since accountability is not sufficient, it would have been important to mention other conditions needed to reduce power imbalances and promote effective global governance. However, as when the Overview calls for '[s]upport for institution building ... so that developing countries can establish appropriate policies and incentives',[80] the emphasis is on the mechanisms in developing countries, and less on international institutions and the power imbalances at that level. This holds also for the version in the full report.

The Overview then calls for 'measures to improve equity and voice in access to financial flows directed at supporting efforts to combat environmental degradation'.[81] The measures are related only to the access to financial flows, and omit other relevant global political and legal instruments and norms; and the talk is of environmental degradation, rather than specifically of climate change. The proposals in the report do not strongly address climate change and its causes, and only seek to 'mitigate' some of its effects. The fashionable term 'voice' helps to hide such silences.

In conclusion, the Overview proposes four country-level sets of tools to take its agenda forward: low-emission, climate-resilient strategies; public-private partnerships; climate deal-flow facilities; and coordinated implementation and monitoring, reporting and verification systems.[82] The final proposal is for

---

79   Ibid 14.
80   Ibid 15.
81   Ibid 14.
82   Ibid 15.

a 'high-profile, global Universal Energy Access initiative dedicated to develop clean energy at the country level'.[83]

In effect, the report makes a case for not using the climate change issue to stop development advances in the underdeveloped world. It argues that: development is needed in poor countries; it can be achieved in a sustainable way; equity is both a means and a goal of development; and equity can lead to sustainability (this link is relatively weakly argued). Thus it argues that development, which is necessary for social justice, can be done without big impacts on the environment. However, it falls short in regard to the need to address more direct measures to reduce climate change, in and by the global North, or to question what type of development is good for the South in the medium and long term.

IV FRAMING AND LEXICAL CHOICE

The central question this chapter asks is in which direction does the HDR 2011 move in relation to the schizophrenic stance in the HDR 2007/2008 on climate change. We have established the basis for an answer by examining in detail the arguments presented in its Overview. We now synthesise that analysis by a review of the inclusions and exclusions that we have seen in the HDR 2011; and then complement this by a comparison of its chosen vocabulary with the vocabularies of the HDR 2007/2008 and, by contrast, the WDR 2010 on climate change and development, to see the direction of movement.

Table 2 brings together points from the previous two parts on the contents of the report's Foreword and Overview. Inside the frame of the HDR 2011 are low-income countries and a certain range of permitted issues, such as local participation, local accountability and gender equality, and wholly or largely outside the frame are rich countries, their production and consumption patterns, global relations, global participation and accountability, and the limits to feasible adaptation in poor countries, the relations between adaptive capacity and governance, the problematic impacts of some types of mitigation strategy, and maladaptation. In sum, the HDR 2011 is in general less questioning of the global status quo, in comparison with the HDR 2007/2008. These exclusions and inclusions become visible through repeated close reading that is guided by a checklist of concerns from critical development studies, political ecology and literature from the field of the human dimensions of global environmental change.[84]

83 Ibid.
84 M Bunce, K Brown & S Rosendo 'Policy Misfits, Climate Change and Cross-Scale Vulnerability in Coastal Africa: How Development Projects Undermine Resilience' (2010) 13 *Environmental Science & Policy* 485, 497; H Eakin, A Winkels & J Sendzimir 'Nested Vulnerability: Exploring Cross-Scale Linkages and Vulnerability Teleconnections in Mexican and Vietnamese Coffee Systems' (2009) 12 *Environmental Science & Policy* 398, 412; BA Beymer-Farrisa & TJ Bassett 'The REDD Menace: Resurgent Protectionism in Tanzania's Mangrove Forests' (2012) 22 *Global Environmental Change* 332,341; Barnett & O'Neill (note 39 above) 213.

Table 2: Spotlights and shadows – inclusions and exclusions in the HDR 2011 with reference to the HDR 2007/2008

Theme / location	Inclusions and/or highlighting	Exclusions and/or downgrading
Title	The vague language of 'equity'	The language of human rights
Title	Gains for everybody	Our divided world; countering global divisions
Geographic focus	LDCs, as a focus for analysis and for policy advice	DCs, global relations
	North as a source of finance (Foreword)	North is not otherwise a target for policy advice (p.iii)
	National and local levels in less developed countries, via focus on problems of the use of land, water and forests (p.6)	The need for changes in highly developed countries, via focus on greenhouse gas emissions.
Topic focus	Diverse environmental and poverty issues	Causes and implications of climate change
Topic focus: Power imbalances	Power imbalances at national level, as a policy focus (Foreword)	Power imbalances at global level, as a policy focus
Analysis: global responsibilities for environmental impact	Carbon emissions per country	P.3: the re-allocation of carbon-intensive production to poor countries, while the output is exported to rich countries (p.3).
Analytical focus	'Win-win-win' in terms of generalized objectives	Describing the lives and livelihoods of specific groups of people (as done more in HDR 2007/8)
Evaluative criteria	HDI (even global HDI); national (and even global) averages and aggregates	Human rights, including of the worst-off groups
	Germany and Sweden as examples of win-win-win policies	Germany and Sweden's questionable performance on greenhouse gas emissions; impacts on poorer groups in tropical and subtropical zones (as described in the HDR 2007/8)
	"the planet" and "future generations" (p.ii)	Current generations of people suffering the consequences of climate change and poverty
Grouping of countries	Very high HDI countries *versus* low, medium and high HDI countries (p.3)	Low HDI countries are not considered separately; instead very high HDI countries are separated out

Theme / location	Inclusions and/or highlighting	Exclusions and/or downgrading
Policy focus	Case for not using the climate change issue to stop development advances in the underdeveloped world.	Reducing climate change, especially through measures in the global north.
	Institution building in developing countries (p.15)	International institutions and the power imbalances at that level. Institutional reform in rich countries
	'Rethinking our development model': the global and international levels as spaces for mobilization of finances	Global and international levels as spaces where rethought structures and systems can be designed and implemented.
	Expanding freedoms for the weak	Restricting those freedoms for the well-to-do which damage the weak
	'Mitigating the effects' of climate change, in other words 'adaptation'	Mitigation: greenhouse gas emission reduction. The limits and barriers to adaptation.
	Population limitation in the global South (p.9)	Consumption limitation in the global North
	Transformation in gender roles in the global South (p.9)	High greenhouse gas emissions etc. in those countries where more gender equality has been achieved
	Democratization in the global South (p.10)	Much greater environmental footprints in many countries with strong democracies, implying there are numerous other relevant conditions
	Empowerment of weaker groups in poor countries (p.10)	Power imbalances between countries
	Finances for adaptation or mitigation of effects in developing countries	Maladaptation, the limits to adaptation. barriers to adaptation, contrasting with the HDR 2007/8 which coined the term adaptation apartheid
	LDCs	Rich countries and rich persons

(a)   Lexical analysis[85]

While parts 2 and 3 conducted the content analysis on a paragraph-by-paragraph basis, and Table 2 reviews the findings systematically, to put figures on some of the tendencies can help to increase confidence in our assessment. We will do this by looking at the choices and frequency of use of keywords. We compare the vocabulary of the HDR 2011 to that of its 2007/2008 predecessor and their WDR 2010 competitor. In each case we confine this analysis to the report Overview.

We remarked earlier that 'climate change' is referred to only 22 times in the Overview for the HDR 2011, compared to 157 times in the Overview for the HDR 2007/2008. While allowing for the greater length of the 2007/2008 report and for the 2011 report's coverage of all aspects of environmental sustainability, this marks a significant reduction of priority to climate change in the later report. Undertaking a similar analysis for other keywords, we find that in the new report the rich or developed countries are little mentioned, and that many of the distinctive features of the 2007/2008 report have disappeared, the features that marked it out as displaying a partly different field of attention and set of values than in the World Bank's comparable WDR 2010. The emphasis on children, grandchildren and future generations is gone, the poor and the world's poor are mentioned four times only, in comparison with 38 times in the HDR 2007/2008, humanity is not mentioned, and other terms such as 'social justice', 'human rights', and 'political' have almost disappeared. The words 'adaptation', 'adaptive', and 'adapt', are used only six times, plus once in the Foreword (versus 68 times in the HDR 2011), and 'mitigation' appears only four times and is used in a very ambiguous way, not referring to reduction or cutting of emissions directly. On the other hand, the terms 'equity' and 'sustain' (in all its variants) are widely used. These terms were characteristic of the 2010 WDR too: 'equity' serves as a vaguer substitute for social justice and human rights, and 'sustain' as part of a growth orientation.[86]

Table 3 provides a comparison of this use of key terms in the overviews of the three reports. A number of complications are faced in making the comparisons. First, the reports vary in length. In the case of the texts of the WDR 2010 Overview and the HDR 2007/2008 Overview there is little difference (with reference to the full version for the HDR Overview and after omitting References in the WDR case since the HDR Overview does not include references), but this is not the case with the HDR 2011. Table 3 therefore presents relative word frequencies rather than absolute word counts. Absolute counts are given in an appendix. They are in themselves significant, since not only the relative frequency of reference to a theme but also the absolute number of words devoted to that theme is an indicator of the importance given to it.

Second, the structures of the documents are not identical. The WDR presents a single version of the Overview, while the HDR presents two, one

---

85   The absolute word counts presented in this section exclude the HDRs' Forewords.
86   Gasper et al (note 8 above).

of which is at the start of the full version of the report, and the other of which is presented within the Summary, which is available as a separate file on the HDR website. As mentioned earlier, this version of the Overview provides a self-contained account that also includes some additional materials from elsewhere in the report, notably selected tables, diagrams and sometimes text boxes. The WDR Overview already incorporates such materials. Further, the WDR Overview incorporates a Reference list, unlike the HDR Overview; while in the other direction, the HDR Summary also incorporates the more substantial HDR Foreword and the Table of Contents, as part of the overall message that it wishes to provide in the version that it expects will be the sole encounter with the report for the majority of readers. Table 3 presents the simplest comparison: excluding in each case the Table of Contents, References, Figures and Tables, and separating out the Forewords. Other comparisons can be added, but inclusion of the additional material will not change the very sharp contrasts in vocabulary that the table reveals.

Table 3: The languages of the three report overviews[87]

Term/phrase frequency of usage (per 1000 words)	HDR 2007/8 Overview	HDR 2011 Overview	WDR 2010 Overview
we	4.6 [6.0]	4.8 [4.8]	0.7
children/our children	0.9 [1.0]	1.1 [1.0]	0.1
grandchildren	0.2 [0.3]	0.1 [0.1]	0.0
future generations	1.6 [1.4]	0.4 [0.6]	0.0
the world's poor and future generations	0.5 [0.4]	0.0 [0.0]	0.0
the world's poor/the world's poorest	2.1 [1.9]	0.0 [0.1]	0.0
the poor / the poorest / the world poorest[as a noun; in addition to uses of 'the world's poor']	1.1 [1.2]	0.6 [0.5]	0.5
the world	2.4 [2.7]	1.4 [1.7]	1.0
humanity	0.7 [0.6]	0.0 [0.0]	0.1
human rights	0.8 [0.7]	0.3 [0.2]	0.0
community/communities	1.0 [1.1]	1.7 [1.9]	0.5
the international community	0.2 [0.3]	0.3 [0.4]	0.0
global community	0.01 [0.1]	0.0 [0.0]	0.0
human community	0.02 [0.1]	0.0 [0.0]	0.0
climate change	12.9 [12.6]	3.0 [2.8]	4.8
justice / injustice	0.7 [0.6]	0.3 [0.4]	0.0
equity / equitable	0.2 [0.1]	8.2 [8.4]	1.0
political/politically/politics	1.7 [1.7]	1.1 [1.1]	0.4
'efficiency'/'efficient'/'inefficient'/'inefficiency / inefficiencies	1.6 [1.5]	0.3 [0.2]	3.1
'climate smart'	0.0 [0.0]	0.0 [0.0]	0.6

87 Word frequency including the Foreword is in brackets. The WDR does not include a detailed foreword, only a brief statement from the president of the World Bank, which is not analysed here. Table of Contents, Notes and References are excluded from the frequency counts.

Term/phrase frequency of usage (per 1000 words)	HDR 2007/8 Overview	HDR 2011 Overview	WDR 2010 Overview
threshold/s	0.6 [0.5]	0.1 [0.1]	0.1
catastrophe/s/catastrophic	0.9 [0.9]	0.4 [0.4]	0.5
insurance, insurers, insure	0.2 [0.4]	0.0 [0.0]	1.1
challenge/s/Challenging	1.6 [1.9]	2.1 [2.2]	0.7
can	2.6 [2.5]	3.6 [3.5]	5.1
manage/(mis)management/ mismanaging	0.6 [0.6]	0.6 [0.5]	2.8
rich countries/rich nations/ rich world/ the rich/er	2.1 [2.0]	0.3 [0.2]	0.1
developed country/ies/world	1.4 [1.2]	0.1 [0.1]	0.3
mitigation/mitigating	3.0 [2.9]	0.6 [0.6]	5.7
economic growth[or growth meaning economic growth]	1.1 [0.9]	0.7 [0.7]	1.6
effective/effectiveness	0.2 [0.1]	1.2 [1.1]	1.3
consumption	0.05 [0.04]	0.04 [0.05]	0.13
sustain[in all variants]	0.26 [0.22]	0.57 [0.62]	0.07
tipping points	0.02 [0.01]	0.03 [0.02]	0.01
fight / fighting	0.05 [0.04]	0.00 [0.00]	0.00
financ/e/es/ing/ed/ial	0.26 [0.23]	0.30 [0.32]	0.55
investment/s/investor/s	0.20 [0.21]	0.14 [0.15]	0.19
adapt/adaptation/adaptive	0.56 [0.52]	0.08 [0.09]	0.35
reduce-cut(ing)/ emissions–greenhouse	0.21 [0.19]	0.01 [0.01]	0.12

In most key respects the HDR 2011 lies clearly closer to the language of the WDR 2010 than to the language of the HDR 2007/2008. Not only does it downgrade attention to future generations, the poor, human rights, and the responsibilities of rich countries, it even drops the concept of 'humanity' which was prominent in the 2007/2008 report and which one would expect to find as part of the perspective that distinctively legitimates preparing a HDR in contrast to the more conventional economic analysis in the WDR. In a few cases, the language is unchanged from 2007/2008: the HDR 2011 does not adopt the WDR 2010's heavy use of 'efficient', 'management' and 'can', or its interest in the economic logic of insurance. It is less averse to the term 'political' than is the WDR, and retains a UN style of speaking in terms of 'we' (35 uses, compared to 56 in the HDR 2007/2008 Overview and only 11 in the WDR 2010 Overview) and of appealing for action in the face of global 'challenges'. But it no longer presents these challenges as exceptionally urgent: use of the idea of 'catastrophe' declines, and the term 'threshold' almost disappears, in both cases down to the same frequency as in the WDR 2010 Overview.

V   IMPLICATIONS OF THE REPORT'S FRAMING FOR AFRICA AND SOUTH AFRICA

Downgrading attention to future generations, the poor, human rights, and the responsibilities of rich countries, along with a focus on technocratic proposed win-win-win synergies between environment and (an instrumental view of) equity and human development, will have serious consequences for sub-Saharan Africa, including for South Africa and especially the poorest groups. We suggest that the report's re-framing of sustainability and its weak treatment of the absolute importance of climate change impacts in the region leads to an insufficient and sometimes misleading picture of the problems faced by the most vulnerable groups. The substantial changes in language, in the framing of the problems and in the solutions offered in the HDR 2011, compared to the HDR 2007/2008, blind the reader to existing systemic socio-economic failures and to constraints on an appropriate balance between economic and environmental issues. In this part we unpack the treatment the report as a whole gives to South Africa and pay particular attention to the implications for South Africa's poor.

First, the Overview – the report's key chapter – only mentions African countries when identifying the likely reduction in human development that environmental degradation and increased climate change impacts will bring to the region.[88] Desertification and deforestation are highlighted as particular worries.[89] But beyond some thin references to the importance of gender equity for sub-Saharan Africa, little else is said about what the HDR 2011 perspective offers to Africa's poor.

The report fails to address properly the fact that for African countries sustainability cannot be disassociated from the pressure that climate change impacts pose. Multiple stressors make climate a development challenge that calls for addressing underlying causes of vulnerability rather than glossing over them. In particular, the report downplays socio-economic, political and institutional factors that limit adaptive capacity and pose barriers to adaptation and resilience. Socio-economic, cultural and political vulnerabilities are recognised in climate change research as factors likely to lead to maladaptation, new vulnerabilities and increased inequalities for Africa's poorest sectors.[90] Moreover, climate impacts in the African continent, if average *global* temperatures rise close to or above two degrees, are likely to be much stronger, and frequently devastating. For example the likelihood of drying for southern Africa increases dramatically as average temperatures rise more than 2°C.[91] Agricultural and food production may be seriously compromised leading to deep food insecurity. Changes in disease vectors,

---

88   HDR 2011 (note 11 above) 2.
89   Ibid 5–6.
90   M Boko et al 'Africa Climate Change 2007: Impacts, Adaptation and Vulnerability' in ML Parry, OF Canziani, JP Palutikof, PJ van der Linden & CE Hanson (eds) *Contribution of Working Group II to the Fourth Assessment Report of the Intergovernmental Panel on Climate Change* (2007) 433, 468; G Ziervogel & A Taylor 'Feeling Stressed: Integrating Climate Adaptation with Other Priorities in South Africa' (2008) 50 *Environment* 32, 41.
91   C Williams, R Kniveton, & R Dominic (eds) *African Climate and Climate Change Physical, Social and Political Perspectives* (2011).

increased cases of malnourishment, and diseases related to extreme events will increase pressures in already weak health systems.[92] The likelihood of conflicts and ethnic tensions due to declining access to resources will increase, jeopardising the prospect of achieving sustainable and peaceful futures.[93] Forced migration due to environmental pressures and changes in livelihoods patterns are also well known consequences of climate impacts, a theme that was well explored in the 2007/2008 report.[94] Also absent in the 2011 report are the problems in the increasingly overcrowded cities of the developing world. And last, the report fails to consider high-end temperature scenarios. Authoritative research shows the impacts of three or four degrees average temperature increase and associated changes in precipitation patterns in the African continent as being dramatic for its socio-economic systems.[95] The report's talk of win-win-win strategies leads to complacency rather than preparedness.

The treatment of less developed countries is in general skewed towards an excessively optimistic picture that hides serious risks and reasons for concern that were documented in detail in the Fourth IPCC Report.[96] With the exception of a single reference to the negative effects of extreme events,[97] the HDR 2011 also fails to take into account the impacts of extreme events in the region. These are well documented in the IPCC Special Report on Extreme Events (SREX), whose summary for policy-makers was released prior to the HDR, and in the report on disaster risk also published early in 2011.[98] Thus the overall framing of the options for a sustainable future in Africa is neither revealing of specific negative conditions attached to low levels of human development, nor sufficiently cautious given the very high risks to food, health and water security, and disaster preparedness and reconstruction that are documented in the SREX, and the substantial changes in precipitation and other likely impacts related to high-end scenarios.

An analysis of references to the region in the overall report unveils an uneven treatment of the structural problems responsible for high inequalities and persisting poverty in Africa. References to South Africa appear mainly in relation to positive synergies and winning strategies for the environment, equity and human development. There are zero references to the continent in the chapter that discusses the limits to human development, although much

92   A Costello et al 'Managing the Health Effects of Climate Change' (2009) 373 *Lancet* 1693, 1733.
93   C Hendrix & S Glaser 'Trends and Triggers: Climate, Climate Change and Civil Conflict in sub-Saharan Africa' (2007) 26 *Political Geography* 695, 715; C Devitt & R Tol 'Civil War, Climate Change, and Development: A Scenario Study for sub-Saharan Africa' (2012) 49 *J of Peace Research* 129, 145.
94   O Brown, A Hammill & R Mcleman 'Climate Change as the "New" Security Threat: Implications for Africa' (2007) 83 *Int Affairs* 1141, 1154.
95   For example, R James & R Washington 'Changes in African Temperature and Precipitation Associated with Degrees of Global Warming' (2012) *Climatic Change*.
96   IPCC 'Climate Change 2007 Synthesis Report. Contribution of Working Groups I, II and III to the Fourth Assessment Report of the Intergovernmental Panel on Climate Change' [Core Writing Team; RK Pachauri & A Reisinger (eds)] (2007).
97   HDR 2011 (note 11 above) 9.
98   IPCC 'Managing the Risks of Extreme Events and Disasters to Advance Climate Change Adaptation (SREX)' (2011); UNISDR 'Disaster Risk Reduction in the United Nations' (2011).

climate change scholarship has shown how poverty conditions and low levels of human development are central factors for vulnerability to environmental change and of paramount importance in limiting adaptation options. While references to South Africa's particular challenges for the poor are absent in the chapters dedicated to poverty or the limits to human development (chapters 1 and 3), South Africa figures in chapters dedicated to success stories and those focused on positive synergies (as in chapters 4 and 5).

Most references to South Africa present an unrealistic picture that hides structural conditions and inequalities that act as barriers and limits to resilience and sustainability. The report talks about expanding access to energy through massive investments in coal-fired plants, such as the World Bank's US$3,75-billion loan to build South Africa's Medupi coal plant, and mentions the 'concerns about greenhouse gas emissions and environmental degradation as well as carbon lock-in when the longevity of infrastructure prolongs the use of obsolete technologies'.[99] But it also presumes such investments will lead to access to energy for poor people, assuming that lack of access to energy in South Africa is only related to availability and not to distributional biases. As Patrick Bond details, there has already been a sharp increase in electricity prices to pay for the building of this plant, in a country where the vast majority of poor people can hardly pay existing prices.[100] Eskom, South Africa's main electricity company, subsidises huge amounts of energy supplied to mining corporations, who pay the lowest energy prices in the world. Expansion along existing lines does not increase jobs, gives no special treatment or effort to assure access to energy for the poor, and the environmental pollution associated with the building and functioning of these coal-fired plants systematically harms poor people.[101]

The report advocates a:

> high-profile, global Universal Energy Access Initiative with advocacy and awareness and dedicated support to developing clean energy at the country level. Such an initiative could kick start efforts to shift from incremental to transformative change.[102]

These claims are decontextualized from existing conditions that prevent access to energy and other basic needs and that act as barriers. The use of the term transformative is thus misplaced, since in the literature it refers to drastic changes to the conditions that produce and perpetuate exclusion.[103]

Similarly, the report presents South Africa as a successful example in integrating the goals of social protection, climate change adaptation and disaster risk reduction in relationship to water management. South Africa's Working for Water Programme, the report argues, includes 'an environmental component, increased stream flows and water availability, improved land

---

99   HDR 2011 (note 11 above) 68.
100 P Bond *Politics of Climate Justice: Paralysis Above, Movement Below* (2011).
101 N Bassey To *Cook a Continent: Destructive Extraction and the Climate Crises in Africa* (2012);
     D Hallowes *Toxic Futures: South Africa in the Crises of Energy, Environment and Capital* (2011).
102 HDR 2011 (note 11 above) 15.
103 K O'Brien 'Global Environmental Change II: From Adaptation to Deliberate Transformation'
     (2012) 36 *Progress in Geography* 667, 676; M Pelling *From Resilience to Transformation* (2011).

productivity and biodiversity in some ecologically sensitive areas'.[104] Even if this programme is an excellent example of conservation that at the same time provides jobs for poorer groups, the absence of a discussion of the deep-rooted problems of lack of access to water and the environmental injustices that plague millions of poor South Africans leads to an unrealistic picture of the feasibility of a sustainable future.[105]

The HDR 2011's focus on win-win-win strategies fails to acknowledge existing inequalities in access to energy and other basic needs, and the forces that generate and perpetuate them, while it highlights the very small advances through fiscal reforms that tax some environmental services.[106] The report further highlights South Africa's claimed success in enforcing environmental rights, whereas the chapter by Dugard and Alcaro in this book documents poor performance of the legal system in relation to the environment.[107] We find a particular mention of how the Constitution of the Republic of South Africa of 1996 guarantees the right of access to any information.[108] But the report remains silent on the underlying processes that reduce freedom and capacity of individuals and groups to respond to information, including illness, illiteracy, innumeracy, and fear for personal safety.

The HDR 2011 treatment of models for rethinking development says next to nothing about the concerns expressed by many civil society organisations about the misuse of the concept 'green economies' in ways that undermine rather than promote sustainable development. Key concerns include the abuse of biofuels, land grabbing, and negative effects of much economic activity labelled as 'green', more commodification of nature, and even more conditionalities and new forms of protectionism.[109]

In short, the depiction of Africa, and in particular South Africa, in this report is not realistic and may lead to complacency and poor national policy-making that remains blind to structural problems and power imbalances responsible for limiting freedoms and capabilities of millions of poor South Africans. It is a different treatment of the problems and opportunities in the region than the one presented in the HDR 2007/2008, and suggests a substantive shift in thinking in the UNDP HDRO.

VI   CONCLUSION

This chapter asked in which direction the thinking on environment and sustainability by UNDP's HDRO has evolved since the HDR 2007/2008, which combined a radical human rights-based diagnosis and critique with largely a conventional economic set of policy proposals, mostly close to those in the WDR 2010. Frame analysis of HDR 2011 indicates its convergence

---

104 HDR 2011 (note 11 above) 78.
105 Bassey (note 101 above); P Bond (note 100 above); Hallowes (note 101 above).
106 HDR 2011 (note 11 above) 8.
107 J Dugard & A Alcaro 'Let's work together: Environmental and socio-economic rights in the courts' (2013) in this volume.
108 HDR 2011 (note 11 above) 87.
109 UNRISD 'Social Dimensions of Green Economy and Sustainable Development' (2011) <http://www.unrisd.org/greeneconomy>.

towards a World Bank perspective: inside the frame are low-income countries and a certain range of permitted issues, including 'extending freedoms', while largely outside the frame are rich countries, global relations, restricting some freedoms, and even, to a surprising and disturbing extent, climate change mitigation. Mitigation of GHG emissions substantially disappears from view and is replaced by talk of 'mitigation of effects'. At the same time, the structural conditions that limit adaptation to the impacts of climate change and prevent resilience, conditions that in addition may lead to maladaptation and generation of new vulnerabilities, are never addressed. Lexical analysis, the comparison of word choices, corroborated this picture.

The HDR 2011 is surprisingly muted on issues of climate change, and does not address key aspects including the need to radically cut GHG emissions. On the contrary, the report takes a conservative position on the issue of cutting emissions, focusing more on expanding access to energy in developing countries in a sustainable way, and neglecting measures to decrease the current impact of unsustainable emissions. Climate change is underemphasised in the report, partially displaced by the concept of environmental sustainability. This takes away urgency from addressing the existing development pathways in the advanced economies primarily responsible for emissions, and thus can be seen as a step backwards. Additionally, the concept of sustainable human development used in the report is formulated in a far weaker way in comparison to the 1987 Bruntland Commission version, by focusing on 'reasonable' efforts instead of outcomes, and by implicitly accepting infringement of future generations' substantive freedoms. In the 2007/2008 report, in contrast, the case for radical action was forcefully made.

The HDR 2011's area of main focus is the relationship between environmental degradation and the promotion of human development, including issues of gender and democratic participation. Not all the links it proposes there are strong, and some are problematic. Emphasis on the political dimensions of climate change and poverty is given only for the local and national levels, leaving invisible the issues of power at the global level. The change of focus in the 2011 report in comparison to the 2007/2008 report is such that the HDR 2011 even – astonishingly – ignores the HDR 2007/2008 when mapping earlier contributions in HDRs on environmental sustainability. Listed on page 14 of the full report are the HDR 1990, the HDR 1994 and the HDR 2010 that emphasised the links between sustainability and human development. Extraordinarily, unmentioned in this brief history is the HDR 2007/2008 devoted to climate change.

Overall the HDR 2011 appears a severely diluted successor to the HDR 2007/2008, and much closer in perspective to the World Bank. This resolution of the 2007/2008 report's schizophrenia is consistent with the backgrounds of the staff who were in charge for 2011. The 2007/2008 report was led by the then head of the HDRO, Kevin Watkins, a political economist who had worked for 20 years in research and programming in human rights-oriented non-governmental development organisations, in particular the Catholic Institute for International Relations and Oxfam UK. Watkins left HDRO in 2008. While the HDRO has editorial independence within UNDP, the head of the office is appointed by the UNDP administrator, in this case Kemal Dervis, who had earlier worked for 24 years in the World Bank. The new head was Jeni Klugman, an Australian economist who moved to the post after 16 years in the World Bank. She was lead author for the 2011 report. Also prominent in the 2011 report team was the HDRO head of research, Francisco R Rodriguez, a Venezuelan economist on leave from Wesleyan University in the US, who took up the HDRO post in 2008 and left for the Bank of America Merrill Lynch in 2011.[110] Klugman returned to the World Bank in 2011. Even some sections in the World Bank now appear bolder than was the HDR 2011, as can be seen in one recent report.[111]

The disappearance of a human rights-based approach from the HDRO work on sustainability is not an incidental detail in this story. A rights-based approach stresses the fundamental importance of basic dignity for all, as something that is not to be traded-off against more consumption benefits for the already affluent. This means that equity is to be seen as an issue with intrinsic importance, not only instrumental. Human rights-based approaches insist on as far as possible specifying and institutionalising systems of obligations to respect and promote basic rights. And they systematically track the causal chains behind rights violations and failures or inabilities to act on obligations, rather than turning one's face away from matters that might be embarrassing for powers-that-be. The deficiencies of the HDR 2011 in all these respects reflects its failure to build on the HDR 2007/2008's rights-based problem diagnosis and to extend it into a rights-based, or at least rights-inspired, approach to policy design. We concluded our analysis of the HDR 2007/2008 with the suggestion that perhaps a different, more widely consultative, mode of report writing could have led to overcoming the gap between its stated values and its policy orientation.[112] The present analysis of the HDR 2011 may suggest that the spirit in which the HDRO was created, to be an independent think-tank able to boldly move forward the debates on development strategy and development cooperation, is now in jeopardy.

---

110   See <http://caracaschronicles.com/2011/10/21/why-chavez-will-win/>.
111   World Bank 'Turn Down the Heat: Why a 4°C Warmer World Must be Avoided' (2012).
112   Gasper et al (note 8 above).

APPENDIX: USAGE OF KEYWORDS IN THE THREE REPORTS' OVERVIEWS[113]

Term/phrase Absolute occurrence	HDR 2007/8 Overview	HDR 2011 Overview	WDR 2010 Overview
we	56 [28]	35 [4]	11
children / our children	11 [3]	8 [0]	2
grandchildren	3 [1]	1 [0]	0
future generations	19 [0]	3 [2]	0
the world's poor and future generations	6 [0]	0	0
the world's poor/the world's poorest	25 [1]	0 [1]	0
the poor / the poorest / the world poorest [as a noun; in addition to uses of 'the world's poor']	13 [4]	4 [0]	8
the world	29 [8]	10 [4]	15
humanity	8 [0]	0 [0]	1
human rights	10 [0]	2 [0]	0
community/communities	12 [3]	12 [3]	8
the international community	2 [2]	2 [1]	0
global community	1 [0]	0 [0]	0
human community	2 [0]	0 [0]	0
climate change	157 [18]	22 [1]	71
justice / injustice	8 [0]	2 [1]	0
equity / equitable	2 [0]	59 [9]	15
political/politically/politics	21 [2]	8 [1]	6
'efficiency'/'efficient'/'inefficient' /'inefficiency / inefficiencies	20 [1]	2 [0]	46
'climate smart'	0 [0]	0 [0]	9
threshold/s	7 [0]	1 [0]	1
catastrophe/s/catastrophic	11 [2]	3 [0]	8
insurance, insurers, insure	3 [3]	0 [0]	16
challenge/s/challenging	20 [6]	15 [3]	11
can	32 [3]	26 [2]	75
manage/(mis)management/ mismanaging	7 [1]	4 [0]	42
rich countries/rich nations/ rich world/ the rich/er	25 [3]	2 [0]	1
developed country/ies/world	17 [0]	1 [0]	4
mitigation/mitigating	37 [3]	4 [1]	84
economic growth[or growth meaning economic growth]	13 [0]	5 [1]	24
effective/effectiveness	2 [0]	9 [0]	19
consumption	6 [0]	3 [1]	19
sustain[in all variants]	31 [0]	41 [9]	11

113 The word counts do not include table of contents, notes and references. We present the word counts of the Forewords in brackets. The WDR does not include a foreword.

Term/phrase Absolute occurrence	HDR 2007/8 Overview	HDR 2011 Overview	WDR 2010 Overview
tipping points	2 [0]	2 [0]	1
fight / fighting	6 [0]	0 [0]	0
financ/e/es/ing/ed/ial	32 [0]	22 [4]	81
investment/s/investor/s	24 [5]	10 [2]	28
adapt/adaptation/adaptive	68 [5]	6 [1]	52
reduce-cut(ing)/ emissions-greenhouse	25 [2]	1 [0]	17

Total word count for each report[114]

HDR 2007/8 Overview	HDR 2007/2008 Overview [Including foreword]	HDR 2011 Overview	HDR 2011 Overview [Including foreword]	WDR 2010 Overview
12,156	13,908	7,232	8,090	14,772

114 Excluding notes, table of contents, references, figures and tables.

SITUATED RESILIENCE: REFRAMING VULNERABILITY AND SECURITY IN THE CONTEXT OF CLIMATE CHANGE

Petra Tschakert and Nancy Tuana

I Introduction

Recent contributions to the climate change debate challenge the dominance of an earth-system driven perspective, highlighting instead the human dimensions that shape vulnerability and adaptation to both the positive and negative impacts of climate change, especially among poor and marginalized populations.[1,2] Particular attention has been paid to multiple stressors, livelihood decision making, power differentials, development, governance, and, more recently, transformation.[3,4]

Within the field of climate change adaptation, two distinct yet complementary concepts have shaped academic discourses and policy making over the last decade: vulnerability and resilience. While the concept of vulnerability points toward critical factors that make individuals, groups, sectors, and regions vulnerable to the impacts of climate change and co-existing other stressors, the concept of resilience stresses key elements that allow people and systems to withstand, recover from, and anticipate climatic and other shifts and disturbances. In order to tackle crucial questions regarding livelihood decision making, governance, and successful adaptation, a closer look at these two concepts is needed. We are particularly interested in examining whether the existing dichotomy between the two approaches constitutes an advantage or a hindrance to understanding relational interdependencies.

Underlying discussions about vulnerability and resilience are concerns with justice to ensure that the harms of climate change are understood and addressed, to be attentive both to current individuals and distant others, as well as relations between peoples and places. One response has been to advocate for a human rights or security framework rather than an economic

---

1   K O'Brien, AL St. Clair & B Kristoffersen *Climate Change, Ethics and Human Security* (2010).
2   AL St. Clair 'Global poverty and climate change: Towards the responsibility to protect' in O'Brien et al (note 1 above) 180.
3   M Pelling *Adaptation to Climate Change: From Resilience to Transformation* (2011).
4   RW Kates, WR Travis & TJ Wilbanks 'Transformational adaptation when incremental adaptations to climate change are insufficient' (2012) 109(19) *PNAS* 7156.

framework for managing the risks of climate change and designing effective response mechanisms to climate impacts.[5,6,7,8]

We argue that debates about vulnerability and resilience, as well as human security, are impeded by an inadequate appreciation of the inherent relationality between peoples and between peoples and places. A more adequate approach to both domains requires a relational ontology which, we contend, will transform all three frameworks. We focus initially on the vulnerability/resilience frameworks to clarify the nature and necessity of a relational ontology, and then demonstrate how it requires not only a transformation of current accounts of resilience and vulnerability, but even the very notion of human security.

II   VULNERABILITY AND RESILIENCE

In a recent review article on possible convergence and synergies between vulnerability and resilience approaches, Miller et al assess differences and similarities regarding epistemological and theoretical traditions, methodologies, and practical applications.[9] While the concept of vulnerability is strongly rooted in hazard studies, disaster risk reduction, and work on food security and sustainable livelihoods, the resilience perspective originates from complex social-ecological systems thinking, with a strong natural science influence. Both approaches are concerned with systemic changes, although a vulnerability lens emphasizes values, agency, assets, and power as the most critical actor-oriented determinants or drivers of change, while the resilience approach is more system-focused, attempting to understand coupled dynamics between social and ecological systems, critical thresholds, and feedbacks. Another distinguishing feature is the resilience community's interest in both slow and fast drivers of change, typically over long time frames, whereas vulnerability studies give preference to shorter temporal scales.

One of the major points of contention is the differential treatment of agency and power. Agency is seen as a core strength of the vulnerability approach while the same analytical concept has remained weak in resilience studies.[10] Additionally, Cote and Nightingale[11] argue that power relations, cultural values, and other normative factors integral to the development and functioning of social-ecological systems have been inadequately captured in resilience studies, reflecting an overemphasis on physical shocks and disturbances to the detriment of social and political change. Although recently

5   S Humphreys & M Robinson *Climate Change and Human Rights* (2010).
6   UNDP United Nations Development Program (1994) *New Dimensions of Human Security.*
7   O'Brien et al (note 1 above).
8   JP Burgess *The Routledge Handbook of New Security Studies* (2010).
9   F Miller, H Osbahr, E Boyd, F Thomalla, S Bharwani, G Ziervogel, B Walker, J Birkmann, S Van der Leeuw, J Rockström, J Hinkel, T Downing, C Folke, & D Nelson 'Resilience and vulnerability: complementary or conflicting concepts?' (2010) 15(3) *Ecology and Society* 11 <http://www.ecologyandsociety.org/vol15/iss3/art11/>.
10   Ibid.
11   M Cote & A Nightingale 'Resilience thinking meets social theory: Situating change in socio-ecological systems (SES) research' (2012) *Progress in Human Geography*, DOI: 10.1177/0309132511425708.

attention has been devoted to the role of adaptive governance and deliberative and multi-layered institutions in addressing questions of what is desirable and resilient and for whom, resilience analyses have remained limited to key structural attributes (flexibility, diversity, and cross-scale connectivity) of institutional arrangements rather than unpacking the decision-making processes and relations that shape these structures. Subjective identities and affective relationships that determine what power is exercised, by whom and over whom, for instance, through class, gender, and ethnicity, constitute crucial elements of dynamic and situated processes;[12] yet, they are typically not part of resilience models.

More forcefully, Cannon and Müller-Mahn[13] express caution regarding the growing dominance of resilience studies, arguing that the intrinsically power-laden connotation of vulnerability is being removed from the heart of climate change debates, thereby de-politicising the very causes that put vulnerable people at risk. They also critique the common assumption that decision making is self-regulating, hence 'rational', further undermining critical reflection on what is right, wrong, sustainable, and fair in negotiating climate responses. Moreover, since many people and livelihoods currently do not experience a resilient condition, the authors claim that vulnerability constitutes a 'more valid' concept as it reflects present and possible future social constructions.[14]

Such a critique is offset by the recognition that the vulnerability concept carries an inherently negative connotation. The labelling of certain groups or regions as vulnerable itself may constitute a type of stigmatization likely to exacerbate marginalisation and, consequently, undercut community agency, autonomy, and just and long-term adaptation.[15] Resilience provides a constructive counter-discourse to narratives of vulnerability that typically describe poor and marginalised populations as passive victims of global changes, including climate change. Unlike vulnerability, a resilience approach provides the necessary discursive and material space for recognising and building adaptive capacity. We concur that a resilience framework can be empowering by providing hope in otherwise overwhelming contexts of uncertainty and risk. However, we argue that debates between resilience and vulnerability frameworks flounder on a series of false dichotomies that are neither necessary nor desirable.

Despite the above noted tensions, a certain confluence between resilience and vulnerability approaches seems to be taking place. Growing emphasis on cross-scalar interactions and processes that allow or hinder transformation in coupled social-ecological systems are likely to bring the two fields closer together, both theoretically and practically.[16] This will require a purposeful shift from 'output-directed to process-oriented research that sees knowledge

---

12   Ibid.
13   T Cannon & D Müller-Mahn 'Vulnerability, resilience and development discourses in context of climate change' (2010) 55(3) *Natural Hazards* 621.
14   Ibid.
15   Ibid.
16   Miller et al (note 9 above).

as co-produced'[17], with particular attention to dynamics that facilitate social/ collective learning, reflection, and planning. Urgently needed for this shift to happen is a better understanding of how to access and incorporate lived experiences of people vulnerable to climate change, in combination with other socio-economic, cultural, institutional, and political shocks and stressors as well as persistent inequalities, and precisely how to build resilience under complex, dynamic, and uncertain conditions.[18] The authors advocate for 'hybrid or pluralistic approaches' as advancing a more integrated understanding of social-ecological change, despite their admission that 'tensions and obvious differences will no doubt persist'.[19]

We caution against a deceptive reconciliation between vulnerability and resilience approaches that obscures a harmful and unnecessary dichotomy lurking beneath the surface. The persistence of the dichotomous use of being either resilient or vulnerable does not permit a deeper appreciation of intrinsic and dynamic interconnections and interdependencies between people, and between people and the environment. While both frameworks acknowledge feedback loops between humans and the environment, referring to these interactions through the hyphenated social-ecological system,[20,21] the accounts remain limited by the persistence of an ontological division between the two domains.

To overcome this dichotomy, we introduce *situated resilience* as an *analytical lens* that more adequately reflects the interconnections between people and the environment. By drawing attention to existing agency as well as limits to adaptation, embedded in spaces of unequal power structures, inequalities, and marginalisation, we propose this practical entry point to understand crucial interdependencies across scales. These range from individuals to institutions and include cross-scalar governance challenges as well as feedback mechanisms across the entire socio-natural realm. We argue that this conception provides a more effective basis for an understanding of human security in the climate change debate that moves away from predominantly individualistic or legalistic solutions. By focusing on situated resilience, we can create the necessary space to examine values and preferences, reconceptualise security, and encourage flexible, forward-looking planning and decision making in the face of climatic and other changes and uncertainties.

III   RESILIENCE AND TRANSFORMATION

Instead of abandoning the concept of vulnerability or of resilience, or trying to argue for the superiority of one framework, we advocate maintaining both

---

17   Ibid 15.
18   Miller et al (note 9 above).
19   Ibid 17.
20   Cote & Nightingale (note 11 above).
21   C Folke, SR Carpenter, B Walker, M Scheffer, T Chapin & J Rockström 'Resilience thinking: integrating resilience, adaptability and transformability' (2010) 15(4) *Ecology and Society* 20 <http://www.ecologyandsociety.org/vol15/iss4/art20/>.

terms, but with significant modifications. We offer the concept of 'situated resilience' rendered consistent with a relational ontology. But as we illustrate, this revised conception is no longer in tension with, but rather intimately linked to a transformed conception of vulnerability, 'corporeal vulnerability' that avoids the problematic dichotomizing that has polarized debates in this arena. This section first articulates the rationale behind retaining a conception of resilience. It then explores the notion of ethical place-making as a bridge between interdependent proximate and distance places, the foundation for introducing a relational ontology of situated resilience and corporeal vulnerability.

(a) Advantages of a resilience lens for exploring transformation

The most innovative aspects of a resilience lens entail the fundamental role of adaptive capacity, the importance of variable, internal change that shapes social-ecological system (SES) dynamics, and the forward-looking perspective that is adopted to embrace unpredictability and change.[22] The notion of adaptability captures the degree to which a SES is capable of self-organisation and the degree to which it can build and increase capacity for learning and adaptation.[23] Moreover, it denotes the capacity for innovation, the ability to learn from mistakes, as well as the capability of dealing with change, adjusting responses to both external and internal change processes, and anticipating the worst and preparing for it. In the context of climate change adaptation, it allows for a specific focus on capacities to learn how to deal and live with change, an ability to embrace change through simple adjustments or more radical, structural transformation rather than to preserve or return to a certain status quo that is presumed to be the desired/desirable norm.

Most importantly, a resilience perspective explicitly deals with transformability, or the ability to create a 'fundamentally new system when the ecological, economic, or social structures make the existing system untenable'.[24] This intrinsic conception of transformative processes is in stark contrast to the narrow engineering notion of resilience that encapsulates a return to initial conditions after disturbance. Learning through experimentation at small scales is essential for shaping transformational change and resilience at larger scales, through cross-scale interactions, innovative network configurations, and the purposeful incorporation of uncertainty and surprise.[25] Understanding system dynamics, interconnections, and feedbacks, including complex social and political processes, have been noted as fundamental for anticipating, adapting to, and managing change.[26]

What do these conceptual advances mean in practice? On-the-ground examples from community-based climate change adaptation projects

---

22 Cote & Nightingale (note 11 above).
23 SR Carpenter, BH Walker, JM Anderies & N Abel 'From metaphor to measurement: Resilience of what to what?' (2001) 4 *Ecosystems* 765.
24 Folke et al 2010 (note 21 above) 3.
25 Folke et al (note 21 above).
26 Miller et al (note 9 above).

underscore the vital relevance of collective and anticipatory learning processes in enhancing adaptive capacity and resilience among resource-poor rural and urban decision-makers. For instance, Fazey et al demonstrate how researchers, community members, NGO workers, and local to regional institutions become 'co-learners' in joint assessment and planning processes in the Salomon Islands.[27] At the core is the recognition of inclusivity and participation in decision-making processes based on negotiated understandings of drivers and trajectories of change, continuous and critical reflection, capacity for dialogue and problem solving, and local ownership and responsibility over named solutions.

In addition to collective learning, a recent widening of the resilience concept illustrates how social capital, vision, leadership, adaptive governance, and the role of institutions for providing access to socio-economic and environmental data and promoting long-term planning provide the necessary discursive arenas for addressing when and for whom possible transformations may be desirable or not.[28] Slow variables such as identity, values, and worldviews are seen as both facilitating and constraining adaptive capacity/capability, and are often dependent on differential power dynamics. Agency, ethics, and governance are increasingly recognised as possible trigger points (critical thresholds) for transformations.[29] Thus, this focus on transformability raises essential questions related to the levels of risk, vulnerability, and loss that may be acceptable. This purposeful inclusion of 'difficult' social, cultural, political, and institutional dimensions into applied resilience thinking – difficult in the sense of integrating them into complex system modelling – provides some of the initial modification of this conception required for it to be compatible with a relational ontology.

(b)   Ethics of place making

Human geographers in particular have been reflecting on the spatial tensions within relational, often urban imaginaries within the broad context of transformation. North[30], for instance, in the context of transition towns (TT) and transition culture (TC), a community-based movement to create places resilient to the threats of peak oil and climate change with origin in the UK, stresses that 'localities, like all places, are differentiated by class, gender and a range of other local oppressions',[31] meaning that a romantisation of TT as harmonious bubbles of equality would be misplaced. He further evokes a spatially wide-ranging notion of justice by requiring intentional localization to tackle processes of uneven development through redistribution of resource endowments, revised international trading rules,

---

27   I Fazey, M Kesby, A Evely, I Latham, D Wagatora, JE Hagasua, MS Reed & M Christie 'A three-tiered approach to participatory vulnerability assessment in the Solomon Islands' (2010) 20 *Global Environmental Change*-Human and Policy Dimensions 713.

28   Miller et al (note 9 above).

29   Folke et al (note 21 above).

30   P North 'Eco-localisation as a progressive response to peak oil and climate change – A sympathetic critique' (2010) 41(4) *Geoforum* 585.

31   Ibid 592.

and the encouragement of endogenous economies. Mason and Whitehead,[32] having been insiders to the TC in the UK, argue that while the movement recognizes the need for 'social justice at a distance' and 'directly obligated care',[33] there are real practical challenges when it comes to ethical place-making:[34]

> In one instance, the movement is characterized by a commitment to localism that could undermine its broad relational sense of care, but tends to support a strong sense of local empowerment and focus in changing the nature of a place. At the same time, however, an ethical commitment of inclusion towards a potentially unlimited local constituency of groups often makes it difficult to develop the oppositional energy that is actually needed to change the nature of a place.

Hence, how can we best apprehend the notion of relational space? Doreen Massey, a Marxist geographer, argues for a 'politics of place beyond place'.[35,36] What she means is a sense of connectedness between the reproduction of life in a certain place and its contribution to the persistence of poverty, inequality, or human rights violations in other places, and the need to rectify associated injustices. Her relational ethics of place is intrinsically political as it allows us to question the origin and trajectories of a certain place, and the political powers in place that shape these trajectories.[37] Massey views understanding and acknowledging the interdependencies between places – which includes people and the environment – as a prerequisite for developing a new ethics of place. For such a new ethics of place to emerge, we have to critically assess the nature of relations that sustain the interdependencies, and evaluate if they are fair, exploitative, sustainable, or unsustainable; it is through this understanding that we are able to develop an ethics of responsibility and care for distant others, and rectify spatially connected injustices.[38]

This implies a 'careful articulation of the relational fabric of a place' (for instance in the calculation of food miles)[39] or the relational integration of space and place by thinking about one's locally-based yet global responsibility. The by-product of climate change, according to Massey, is that it 'unseats' what is typically considered as local, it identifies the 'changeability' of a place, and its 'inevitable openness', all of which make a place 'unbounded'.[40] Hence, when thinking relationally, this particular notion of unbounded places opens an important door to envisioning the socio-natural as inextricably linked, beyond feedbacks in coupled social-ecological systems, in fact close to the notions of porosity and situated resilience that we introduce below.

---

32   K Mason & M Whitehead 'Transition Urbanism and the Contested Politics of Ethical Place Making' (2012) 44(2) *Antipode* 493.
33   Ibid 507.
34   Ibid 511.
35   D Massey *For Space* (2005).
36   D Massey *World City* (2007).
37   D Massey, S Bond & D Featherstone 'The Possibilities of a Politics of Place Beyond Place? A Conversation with Doreen Massey' (2009) 125(3-4) *Scottish Geographical Journal* 401.
38   Mason & Whitehead (note 32 above).
39   Ibid 508.
40   Massey et al (note 37 above) 412-413.

IV   RELATIONAL ONTOLOGY AND SITUATED RESILIENCE

Reflecting on a relational ethics of place evokes the possibility of a relational ethics of resilience. It is analogous to teleconnected vulnerability,[41,42] suggesting the existence of linked and nested interactions and feedbacks related to environmental change, economic markets, and flows of resources and information between distant people and places. Such teleconnections also entail feedbacks of local adaptive responses shaping the manifestation of global environmental and socio-economic change and household choices in distinct and often quite distant geographic contexts. Sustainability (and, by analogy, resilience) and vulnerability of specific individuals and communities are not geographically bounded but connected across scale.

Thinking about places beyond place, the connectedness across spatial and temporal distances, and the necessary ingredients for better (more just) place-building practices brings to the fore, more clearly, the problematic tension at the core of dominant approaches to social-ecological change, namely which side of the 'social' / 'ecological' divide theorists emphasize. Alas, efforts to catalyse convergence between these frameworks,[43] are very likely to run aground due to an untenable, but firmly held, ontology that conceptualises these domains as discrete and separable. We argue that a more adequate ontology, one arising from the inextricable interrelations between humans, and between humans and the world they are of and in is required to successfully accomplish this goal. Such a claim is in line with recent advances in critical human geography that bring together relational approaches to scale, more-than-human ontologies, and unpredictability in encounters to reframe climate change and climate change responses.[44,45,46] We thus advocate a *relational ontology* that avoids the counterproductive either/or of the social and the ecological, and a series of concomitant dichotomies.

(a)   Relational ontology

The ontological commitments that are embedded in vulnerability/resilience frameworks are well entrenched in Western conceptions of the world and hinge around a series of dichotomies that frame our conception of humans and nature, and our theories of their interrelations. Divisions between mind and matter; subject and object; subjectivity and objectivity have structured not only philosophical theories, but Western theories and practices of science.

---

41   WH Adger, H Eakin & A Winkels 'Nested and teleconnected vulnerabilities to environmental change' (2009) 7 *Frontiers in Ecology and the Environment* 150.
42   H Eakin, A Winkels & J Sendzimir 'Nested vulnerability: exploring cross-scale linkages and vulnerability teleconnections in Mexican and Vietnamese coffee systems' (2009) 12(4) *Environmental Science & Policy* 398.
43   Miller et al (note 9 above).
44   L Head & C Gibson 'Becoming differently modern: Geographic contributions to a generative climate politics' (2012) *Progress in Human Geography*, DOI: 10.1177/0309132512438162.
45   C Brace & H Geoghegan 'Human geographies of climate change: Landscapes, temporality, and lay knowledges' (2011) 35(3) *Progress in Human Geography* 284.
46   P Tschakert & AL St. Clair 'Conditions for transformative change: The role of responsibility, solidarity, and care in climate change research' (under review) *Hypatia*.

Whitehead, a vociferous champion of refusing these dichotomies, provided the following account of the ontology he opposed:[47]

> [One] way of phrasing this theory which I am arguing against is to bifurcate nature into two divisions, namely into the nature apprehended in awareness and the nature which is the cause of awareness. The nature which is in fact apprehended in awareness holds within it the greenness of the trees, the song of the birds, the warmth of the sun, the hardness of the chairs, and the feel of the velvet. The nature which is the cause of awareness is the conjectured system of molecules and electrons which so affects the mind as to produce the awareness of apparent nature.

According to Whitehead, this ontology has resulted in an academic divide between scientific research, on the one hand, and social scientific and humanities research on the other, in which the former is focused on the material world of nature and the latter attends to the domain of the beliefs and experiences of the world's subjects, typically assumed to be limited to humans. The dichotomy is a sharp one: the natural world is posited as existing apart from the experiences of thinking subjects, devoid of both agency and affectivity. On such an account, science becomes a domain of objective knowledge, placing it apart from the subjective realm of human meanings and experiences which, in turn, becomes the domain of the social sciences and humanities.

We argue that the unwitting internalisation of a subject/object ontology is a barrier to current efforts to support 'convergence' between vulnerability approaches, which focus on actor-oriented accounts with their concurrent emphasis on values, interests, agency, and knowledges, and resilience approaches, which traditionally focus on coupled systems and feedbacks. Efforts to encourage 'pluralism' or 'hybridization' will be hindered by the pull of dichotomous habits. What is needed is an appreciation that these are false dichotomies, arising from incomplete and inaccurate understandings of the rich interactions or interrelationality between what is, namely between people and places and between the natural and the cultural. Insights from feminist philosophy help us comprehend the permeability (porosity) and shifting nature of seemingly straightforward division between the human and the natural:

> ... the problem arises from questionable ontological divisions separating the natural from the humanly constructed, the biological from the cultural, genes from their environments, the material from the semiotic. We can make divisions between the biological and the social, as we feminists did with sex and gender, but what we soon discovered is that the divisions are both permeable and shifting, while at the same time deeply entrenched in bodies and practices.[48]

We thus advocate a *relational ontology* as the way to avoid the either/or of the social/natural in and realise their relationality. Key to the appreciation of a relational ontology is understanding how the material and the experiential/conceptual are intertwined and co-emergent. Donna Haraway's phrase

---

47   A N Whitehead *Concept of Nature. The Tarner Lectures Delivered in Trinity College November 1919.* Project Gutenberg (1919/1964) 16 <http://manybooks.net/titles/whiteheada1883518835-8. html.

48   N Tuana 'Viscous Porosity: Witnessing Katrina' in S Alaimo and S Hekman (eds) *Material Feminisms* (2008) 189.

'material-semiotic'[49] as well as Eric Swyngedouw's term 'socionature'[50] are examples of concepts designed to capture the ever-connected relationality between the biological and the cultural.[51] Swyngedouw's conception of socionatural acknowledges that:

> natural or ecological conditions and processes do not operate separately from social processes, and that the actually existing socionatural conditions are always the result of intricate transformations of pre-existing configurations that are themselves inherently natural *and* social.[52]

A relational ontology shifts the debate from scientific accounts of the material world versus social scientific and humanities accounts of the human world, to the continuous and emergent interplay between entities, precluding not only a sharp divide between the cultural and the natural world but also between the human and the natural sciences. Rather than the anthropocentric bias that humans are the only actors or the dichotomy between social and natural systems, a socionatural perspective recognises that there are only socionatural interactions.

To illustrate the co-constitution of the socionatural, consider the people and places of the city of New Orleans in the United States.[53] New Orleans is what it is today because of a series of material-semiotic interactions: an area surrounded by water—the Gulf of Mexico, the Mississippi River, and Lake Pontchartrain; the importance of the Mississippi River for trade and transit; the wetlands surrounding the Delta and the illnesses they breed; the long-term and continuous practices of controlling the waters that surround the city, from the levees that have shifted the course of the Mississippi River and contributed significantly to the loss of about 75 square kilometres of Louisiana coastal wetlands annually; a city that is built on land that is an average of six feet below sea level; the removal of water to create usable land, which in turn causes the land to sink further; a levee system that results in river levels sometimes being significantly higher than street levels; people whose risk perceptions allow them to accept river levels that are 6-10 feet higher than their homes; the poverty and racism that resulted in thousands of New Orleaneans without homes or resources in the wake of Katrina and Rita. The point here is that it is neither effective nor epistemically responsible to try to divide this example, or any other, into the social and the natural aspects, even if we consider them 'coupled'.

---

49   D Haraway *Modest_Witness@Second_Millennium.FemaleMan©_Meets_OncoMouse: Feminism and Technoscience* (1997).

50   E Swyngedouw 'Modernity and hybridity: Nature, *regeneracionismo*, and the production of the Spanish waterscape, 1890-1930' (1999) 89(3) *Annals of the Association of American Geographers* 445.

51   A Escobar (2010) develops the concept of the socionatural world in his work. He defines the term as 'the complexity of relations between the biophysical and human domains...that account for particular configurations of nature and culture, society and nature, landscape and place, as lived-in and deeply historical entities'. However, his account, unlike Swyngedouw's, risks anthrocentrism in continuing to privilege human action. Eg 'socionatural worlds become the result of human actions even if conditioned by particular environments.'

52   Swyngedouw (note 50 above) 445.

53   This account is developed in depth in Tuana (note 48 above).

The ontological inseparability of nature and society demands that our theories and practices, including our understandings and responses to climate change, be transformed by a relational ontology. Such an ontology apprehends the conjoint constitution of nature and society undermining the troubling dichotomy between 'the nature apprehended in awareness and the nature which is the cause of awareness'.[54] A relational ontology also undercuts a series of false dichotomies that trouble debates between resilience and vulnerability theorists including the false divide between facts and values, and between the local and the global.[55,56,57] It requires that we appreciate how deeply *situated* all things are in materiality, values, habits, and interests. It requires a full appreciation of the depth of the interconnections between near and distant others, both spatially and temporally, and grounds an understanding that these others are not only human others. A relational ontology thus requires being *accountable* to the ways we are engaged in material-semiotic practices. Adopting a relational ontology renders meaningless the distinction between a systems approach ('resilience') versus an actor-oriented approach ('vulnerability'), and provides the foundation for a new framework based on an appreciation of the co-constitution of people and places.

(b)   Reciprocal and corporeal vulnerability

Another dichotomy, or at least a value-laden continuum, haunts the vulnerability/resilience debates. Vulnerability, while acknowledged as a relational term, is always framed as a negative state. The higher the risk of harm or injury, the greater the vulnerability. Vulnerability is thus defined as the 'degree to which a system is susceptible to injury, damage or harm' and 'the propensity or predisposition to be adversely affected'.[58]

Adaptive capacity, and often resilience, is defined as 'the combination of the strengths, attributes, and resources available to an individual, community, society, or organization that can be used to prepare for and undertake actions to reduce adverse impacts, moderate harm, or exploit beneficial opportunities'[59] with adaptive capacity seen as inversely correlated to a group's vulnerability. Here we see another divide, in this case with normative connotations, resulting in the common interpretation of vulnerability as resilience's opposite. 'The conventional and tacitly assumed understanding holds that to be vulnerable is simply to be susceptible, exposed, at risk, in danger. In short, it is to be

---

54   Whitehead (note 47 above) 16.
55   SG Bunker *Underdeveloping the Amazon* (1985).
56   N Castree & B Braun 'The construction of nature and the nature of construction: analytical and political tools for building survivable futures' in B Braun & N Castree (eds) *Remaking Reality: Nature at the Millennium* (1998).
57   Haraway (note 49 above).
58   'IPCC authors and expert reviewers annex' (2012) in CB Field, V Barros, TF Stocker, D Qin, DJ Dokken, KL Ebi, MD Mastrandrea, KJ Mach, GK Plattner, SK Allen, M Tignor & PM Midgley (eds) *Managing the Risks of Extreme Events and Disasters to Advance Climate Change Adaptation.* A Special Report of Working Groups I and II of the Intergovernmental Panel on Climate Change (IPCC) 545-553.
59   Ibid 556.

somehow weaker, defenseless and dependent, open to harm and injury'.[60] A relational ontology exposes this conception of vulnerability/resilience (bad/ good; to be reduced/to be enhanced) as inadequate and undermines the commonly posited divide between power and vulnerability.

A relational ontology entails an ethics of relation, one in which relationships between selves and others are characterised by *reciprocal vulnerability*. A number of theorists have advocated a 'new bodily ontology' as a key component of a relational ontology.[61,62,63] Here, vulnerability is not framed as a negative, but rather as an openness to the other through which each being's uniqueness emerges. Vulnerability is neither passive nor negative, but rather the ability to affect and be affected. Unlike traditional accounts where only some are vulnerable, *corporeal vulnerability* recognizes the ontological condition of relationality as the precondition for interrelationality. Corporeal vulnerability denotes the openness essential to relationality. This is the precondition for care as well as for injury, for benefit as well as for harm, and for the ability to give care as well as to inflict injury. As Judith Butler[64] explains, in the case of the political domain:[65]

> Mindfulness of this vulnerability can become the basis of claims for non-militaristic political solutions, just as denial of this vulnerability through a fantasy of mastery...can fuel instruments of war. We cannot, however, will away this vulnerability. We must attend to it, even abide by it as we begin to think about what politics might be implied by staying with the thought of corporal vulnerability itself.

Peoples and places are vulnerable to one another because of their *porosity* and inter-relationality. This vulnerability, this openness to emergence through interconnection, is what allows for and supports relationality, and is productive of the specific forms of relations that exist at a particular time and in specific contexts. Whether this productivity is harmful or beneficial is richly situated, determined by historical contexts, biological interconnections, social institutions, ecological dynamics, interests, and aims. It is also the reason that a relational ontology is inherently an *ethical* ontology. Reciprocal vulnerability is the wellspring of the affective and the ethical, and the condition of care and responsibility. Vulnerability is neither inherently good nor bad; it is rather the condition of being in relation to others, including non-human others. Recognition of the openness of peoples and places to both benefits and harms, to what Des Gasper[66] calls 'shared fragility', can motivate ethical

---

60   E Gilson 'Vulnerability, ignorance, and oppression' (2011) 2 *Hypatia: A Journal of Feminist Philosophy* 308, 309-310.
61   J Butler *Giving an Account of Oneself* (2005).
62   J Butler *Frames of War* (2009).
63   A Cavarero *Relating Narratives: Storytelling and Selfhood* (2000).
64   J Butler *Precarious Life: The Powers of Mourning and Violence* (2009) 29.
65   While Butler and Cavarero are arguably the leading contemporary proponents of corporeal vulnerability, their accounts suffer from a failure to sufficiently appreciate that relationality is not limited to exchanges between humans, but also includes the nonhuman. Extending the domain of corporeal vulnerability to all aspects of relationality is thus an essential step missing from these accounts.
66   D Gasper 'Climate change and the language of human security' (2010) ISS Working Paper #505 25 <http://repub.eur.nl/res/pub/19843/wp505.pdf>.

place making and attention to relationships of care and flourishing connected across space and time.

Corporeal vulnerability thus occasions ethical obligations, to strive for relations between people and between people and places such that there is a fair balance between who and what is harmed and who and what benefits from their relational vulnerability. Rather than attempting to eliminate vulnerability and boost flourishing through adaptive capacity or resilience, the aim is to keep them together, in a balance that can be ethically justified. According to Butler, this means that 'the recognition of shared precariousness introduces strong normative commitments of equality and invites a more robust universalizing of rights'.[67] While essential that her account be extended to include non-human vulnerability, it foregrounds the inherently ethical fabric of a relational ontology.

(c) Situated resilience

Corporeal vulnerability displaces the notion of vulnerability being an inherently negative and passive state, but it demands that we examine the particular ways in which people are *situated* and how their material-semiotic situatedness co-constitutes the distribution of benefits and harms. It means that we must attend to the ethics of place in order to understand how ways of living in one situation are interconnected to ways of living – both human and non-human – in other places. The goal is to get away from a binary conception of a person, species, community, ecosystem, sector, or region being *either* vulnerable or resilient.

To demarcate our conception of resilience from more traditional accounts of social or social-ecological resilience, we label our conception *situated resilience*. It plays an important role in thinking about forms of relationality and how they might provide a basis for people and places to respond to changing conditions in ways that keep an ethical balance between who and what is enabled to flourish and who and what is harmed. A conception of situated resilience animated through an understanding of corporeal vulnerability simultaneously reinforces an appreciation of the porosity of our relations to others, both human and non-human, near and distant, along with the ability to absorb shocks, experience and overcome loss, embrace change, connect, reflect, anticipate, and transform. However, to retain this conception of resilience, it is crucial that it be understood in such a way as to avoid reanimating the false dichotomies discussed above.

Our conception of situated resilience is influenced by the work of Donna Haraway[68] in 'Situated Knowledges'. Like us, she urges avoiding binary oppositions and develops a conception of knowledge that is always *located* historically, bodily, culturally, and spatially. According to Haraway, knowledge is *partial*, that is, attentive to the dynamics of interest and power

---

67   Butler (note 62 above) 29.
68   D Haraway 'Situated knowledges: The science question in feminism and the privilege of partial perspective' (1988) 14(3) *Feminist Studies* 575.

on what is known and not known, as well as *accountable*, not only for what is
and is not known, but also what is and is not enacted given those knowledges.
She also extends knowledge from the traditional assumption of knowledge
as limited to humans, to recognise both non-human and non-individual
forms of knowing. Haraway argues for a politics of positioning, in which
we recognize and embrace partial perspective as a model of knowledge that
avoids the relativism-absolutism dichotomy and recognize the inseparability
of the normative from the known. That is, admitted or not, politics and ethics
ground struggles over knowledge projects in the exact, natural, social, and
human sciences'.[69]

**Table 1: A sequence of resilience concepts, from the more narrow interpretation
to the broader, inextricably linked socio-natural context (light grey)
(after Folke[70]).**

Resilience concepts	Characteristics	Focus on	Context
Engineering resilience	Return time, efficiency	Recovery, constancy	Vicinity of a stable equilibrium
Ecological/ ecosystem resilience, social resilience	Buffer capacity, withstand shock, maintain function	Persistence, robustness, landscapes	Multiple equilibria, stability
Social–ecological resilience	Interplay disturbance and reorganization, sustaining and developing	Adaptive capacity, learning, innovation, transformability	Integrated system feedback, cross-scale dynamic interactions
Situated resilience	Material-semiotic relationality, unboundness, permeability, situatedness, partiality, accountability	Relational ethics, webbed connections, interdependencies between people and places, politics of positioning, care, responsibility, reciprocal vulnerability, mutual fragility	Balance between flourishing and harm, ethical place making, historically, bodily, culturally, and spatially located knowledge, deliberate transformative change

We propose situated resilience as an extension to a frequently used typology
of resilience concepts, as understood in the global change/climate change
community (Table 1). Situated resilience, like situated knowledges, is
anchored in the particularity of material-semiotic relationality. It was always
partial and positioned, always resilience for a particular collection of entities
in a particular context. And situated resilience is always power-sensitive—
resilient for whom and why. It recognises that resilience is never final, but
always emergent from webs of relationships. It is not about invulnerability

69   Ibid 587.
70   C Folke 'Resilience: the emergence of a perspective for social-ecological systems analyses'
      (2006) 16(3) *Global Environmental Change* 253.

but about better ways of encouraging relations between peoples, both current and future, and between peoples and places. Situated resilience proceeds from the corporeal vulnerability of the socionatural and is thereby attentive to the inseparable interactions between biophysical and socio-political dimensions of relationality. Our notion of situated resilience is thus a response to the effort by Cote and Nightingale[71] to embed the concept of social resilience in social relations of power, knowledge, and culture.

V   SITUATED RESILIENCE AND THE VERY NOTION OF HUMAN SECURITY

Situated resilience is an inherently normative framework. Through an appreciation of the inextricable interconnections between peoples and between peoples and places, and its responsiveness to corporeal vulnerability, situated resilience entails a relational ethics that is attentive to power and the politics of positioning, as well as to care, accountability, and responsibility. It thus demands a normative rather than an econometric response to climate change. While there have been efforts to embed ethics into climate policy discourse via rights or justice approaches, we argue that such approaches will only be effective if they are based on a relational ontology that avoids reanimating dichotomous habits of ignoring the interconnections between people and between people and places, as well as escaping the temptation to see affect as antithetical to reason. To illustrate what this means, we examine the normative framework we see as most promising in the domain of climate change, human security, and clarify how it must be modified given our analysis.

(a)   Human security in the context of climate change

In his report for the 2005 World Summit, 'In Larger Freedom: Towards development, security and human rights for all', Kofi Annan exhorts that 'humanity will not enjoy security without development, it will not enjoy development without security, and it will not enjoy either without respect for human rights'. The concept of human security was formally introduced through the United Nations Development Program (UNDP 1994), in which the focus of security was shifted from the national to the individual level. 'The world can never be at peace unless people have security in their daily lives' and claims that security for people consists not of security of a nation from external aggression, but rather 'protection from the threat of disease, hunger, unemployment, crime, social conflict, political repression and environmental hazards'. [72]

The human security framework aims to shift the focus of security from militarized protection of the state to protection of individuals from a wide range of threats to their well-being and security. Human security aims to address the basic needs required for individual wellbeing, including water and food security, health security, community/identity security, and security of political freedoms. But it goes beyond needs discourses to emphasize empowerment,

---

71   Cote & Nightingale (note 11 above).
72   UNDP (note 6 above) 23.

protection, and responsibilities. While the UNDP report insisted that human security was a universal concern, 'relevant to people everywhere, in rich nations and poor',[73] it recognizes that what constitutes security is not fixed for all people in all contexts, but is a factor of both individual and social values and contexts. It is also attentive to increasing inequities between peoples, and the institutions and systems that cause them[74]. An often cited goal of a human security framework is freedom from fear and freedom from want.[75]

Human security approaches focus on 'the human species as a whole, and its shared security, insecurity and fragility'.[76] Gasper and Truong[77] argue that a human security perspective provides the basis for a 'joined-up feeling' which animates feelings of sympathy, solidarity, and even responsibility. Its aim is to help everyone, including those who are currently well-off, appreciate that their security is dependent on the security of all people. 'To an awareness of and respect for individuals it adds an understanding of human individuals, the category of human species, and a sensitivity to the specifics of human need, vulnerability and shared insecurity, wherein each affects all'.[78]

While there are many positive features of a human security approach, the framework lacks sufficient appreciation of the depth of relationality required for a fully adequate "joined-up feeling." In the next section, we identify three weaknesses of current human security approaches and provide an alternative conception of security, what we call *socionatural security*, designed to strengthen a security approach using insights from the perspective of a relational ontology.

(b)   Socionatural security

Human security approaches are clearly systems focused, attentive to the interdependence of the components of human security. 'It is inherently an integrative and relational concept that draws attention to present and emerging vulnerability that is generated through dynamic social, political, economic, institutional, cultural and technological conditions and their historical legacies'.[79] While recognising the value of a human security approach, a relational ontology and the associated appreciation of reciprocal vulnerability and situated resilience argues for an even broader conception of security. This conception, what we call *socionatural security*, recognises the integral and indissoluble interconnections between humans and the natural world that they are of and in.

73   Ibid.
74   B Hayward & K O'Brien 'Social contracts in a changing climate: Security of what and for whom?' in O'Brien et al (note 1 above) 199.
75   UNDP (note 6 above) 24.
76   D Gasper The Human Security Approach as a Frame for Considering Ethics of Global Environmental Change (2009) IHDP 2.
77   D Gasper & T-D Truong 'Deepening development ethics through the lenses of caring, gender, and human security' (2008) ISS Working Paper #459.
78   Gasper (note 75 above) 18.
79   K O'Brien, AL St. Clair & B Kristoffersen 'The framing of climate change: Why it matters' in O'Brien et al (note 1 above) 5.

Reciprocal vulnerability, between people and between humans and nature, extends and transforms human security by displacing the focus on humans, a focus that risks obscuring the porosity that links and situates social and natural resilience and well-being. Attention to a relational ontology, we argue, refocuses security discourses to better reflect and appreciate three forms of interconnection that are not sufficiently attended to in contemporary human security discourses. These are: 1) the socio and the natural; 2) reciprocal corporeal vulnerability and situated resilience; and 3) close and distant others.

Figure 1 incorporates all three dimensions of interconnection while highlighting multi-scalar interactions that are always in the process of being made and negotiated, emergent and never closed. Following Massey[80], we argue for an outwardlookingness, and a genuine openness of/toward the future, a space of co-constituent becomings, including surprises, change, experimentation, transformation, and gaps, loose ends, and missing links. We do not want to suggest that all connections are already known, and determined, along one singly future story.

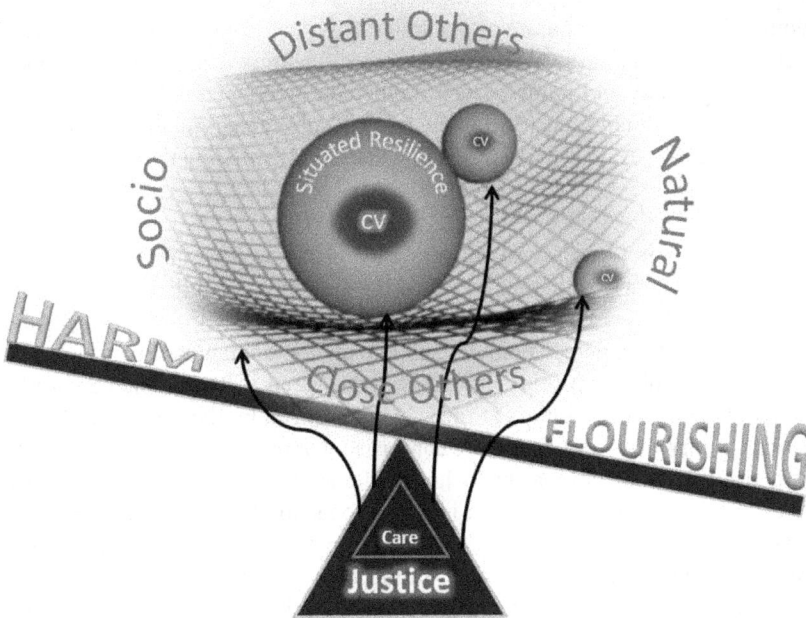

Figure 1: Socionatural security: through a lens of a relational ontology of resilience (coloured bubbles = situated resilience)

80   Massey (note 35 above).

(i) Socionatural

The human security framework is centred on people and their social relations.[81] That focus, unfortunately, obscures the inextricable interconnections between human and environmental security. This is not to say that advocates of a human security perspective do not see the importance of human relations with the environment as a factor in human well-being. The Commission on Human Security, for example, recognizes 'the intrinsic value of the dignity of all human beings in a holistic way that includes their dependency and their relations with the natural environment'.[82] Gasper, to provide a second example, argues that a human security approach aims to emphasise 'human solidarity, stability and prioritization, prudence and enlightened self-interest, sources of richer quality of life, felt security and fulfillment, and ecological interconnection that demands careful stewardship'.[83] Yet, while noted, the reciprocal vulnerability of humans and environments has not been fully developed in current accounts. We suspect the pull of dichotomous thinking and the privileging of the human domain, is at fault, but the re-emphasis of 'human' security over 'socionatural' security obscures appreciation of this important interrelation. As climate change so powerfully illustrates, human flourishing is never separate or separable from environmental or ecosystem flourishing. Given the sway of anthropocentrism, and the metaphysical divide between humans and nature deeply embedded in Western metaphysics, we contend that human security frameworks risk obscuring the mutual porosity of socialnatural systems, which is a key feature of a relational ontology.

(ii) Reciprocal vulnerability

An appreciation of corporeal vulnerability is the well-spring for 'joined-up thinking and joined-up feeling'.[84] While the authors argued that this joined-up thinking and feeling become a cosmopolitan concern, attention to socionatural security and the appreciation of the porosity of human/nature interactions requires that joined-up thinking and feeling extend to the nonhuman world. While this mutual porosity is at times acknowledged by those working from a human security framework, we are concerned that the focus on human security will drive attention primarily to human flourishing and sideline ecosystem flourishing. As just one example, in the introduction to the anthology *Climate Change, Ethics, and Human Security*, O'Brien et al note that 'climate change represents one more factor demonstrating how human activities are altering ecosystems and ecosystem services, which in turn have implications for human wellbeing'.[85] However, in the dozen articles that make up the anthology, the environment and environmental security are muted at best. As we argued above, situated resilience requires full attention to both aspects of flourishing, without creating a problematic divide between the social and the natural.

---

81   O'Brien et al (note 1 above).
82   Ibid 5.
83   Gasper (note 75 above) 18.
84   Gasper & Truong (note 76 above).
85   O'Brien et al (note 1 above) 5.

(iii)   Close and distant others

Human security theorists have consistently appreciated the inter-relationality between people in distant places, and some explicitly recognise the responsibility to protect the security not only of current but also of future generations.[86] An ontology of relationality and an appreciation of porosity between humans and between humans and places provide the metaphysical basis for both intra- and inter-generational justice in the context of security. What a socionatural approach adds in addition to providing this metaphysical basis is an appreciation of the importance of focusing reason and affect upon the types of interactions between people and places that support the flourishing of distant others, including those who are spatially as well as temporally distant, and those which are more likely to cause an imbalance of harm and flourishing.

We view *socionatural security* as an important evolution of the current conception of human security (Table 2). Just as most who advocate a human security approach, viewing it as building on and expanding from a human rights perspective, we similarly advocate socionatural security as a needed evolution of human security approaches.

Table 2: A comparison of human rights, human security, and socionatural security frameworks

Framework	Beneficiary	Goal	Focus
Human Rights	individual/state	basic needs	legal focus
Human Security	the human species as a whole	felt security and well-being	stability of human lives
Socionatural Security	the socionatural system	socionatural flourishing	stability of socionatural interactions

VI   Conclusion: Toward a New Paradigm

We have argued that frameworks such as resilience and vulnerability, designed to address crucial transformations essential to responding to climate-related impacts, have been constrained by intrinsically problematic dichotomies. We contend that an ontological comprehension of the depth of interrelationality has the potential to transform these approaches and liberate them from their current discursive traps. Our main focus has been on how this ontology gives rise to an appreciation of situated resilience, thereby reframing both resilience and vulnerability frameworks. At the same time, we have gestured at how this new ontology amends conceptions of human security to the more adequate notion of socionatural security. This reframing has the potential to lead to new directions in the understandings and practices of transformational communities and ethical place making. While attempting to effectively communicate the radical possibilities for this relational conception, this essay is best understood as a call for research and practice aimed at fully appreciating the value of this new paradigm for responding to climate change.

---

86   St. Clair (note 2 above).

COP-ING OUT

CLIMATE CHANGE, POVERTY AND CLIMATE JUSTICE IN SOUTH AFRICAN MEDIA: THE CASE OF COP17

JILL JOHANNESSEN

I INTRODUCTION

Narratives of climate change are becoming central to development discourses and increasingly frame understandings of global challenges, such as poverty, health, energy, and food security.[1] The relation between climate and poverty is also increasingly of concern to politicians who see these two challenges as inextricably linked. The South African government has stated that climate change challenges must be coupled with socio-economic and development challenges, including the situation of the poor.[2] By hosting the 17th Conference of the Parties (COP17)[3] at the end of 2011, the South African government had an opportunity to reframe climate change as a holistic sustainable development challenge.

COP17 attracted tens of thousands of people hoping to affect political outcomes. It embodied a 'shared' moment in which the different member states mobilised expertise and power, including various activists who were struggling to construct meaning and voice their concerns regarding the negotiations. The mass media is a key channel of communication, mediating between the different groups of actors and competing knowledge claims, framing climate issues for politics and the public, and drawing attention to how to make sense of, and value, the changing world.[4]

Global media monitoring shows that news coverage of climate change has peaked during the yearly United Nations climate summits, with COP15 in Copenhagen (2009) as the largest peak.[5] COP17, by contrast, was not a global media priority,[6] probably due to low expectations and the lack of novelty. However, climate change is a rather new topic for most mainstream media in

---

1   See UNDP 'Human Development Report (HDR) 2010: Fighting Climate Change – Human Solidarity in a Divided World' (2007/2008); World Bank 'World Development Report (WDR) 2010: Development and Climate Change' (2010).
2   See National Climate Change Response White Paper (2011) <http://www.info.gov.za/view/Dyna micAction?pageid=623&myID=315009>.
3   See <http://www.cop17-cmp7durban.com/>.
4   MT Boykoff *Who Speaks for the Climate? Making Sense of Media Reporting on Climate Change* (2011) 3.
5   Boykoff et al track newspaper coverage of climate change or global warming in 50 newspapers across 20 countries and six continents <http://sciencepolicy.colorado.edu/media_coverage/>.
6   W Schreiner & J Bosman 'Coverage Cop-out: Global Media Analysis Points to a Lack of Climate Change Coverage' (2012) 33 *Ecquid Novi: African Journalism Studies* 66.

South Africa and due to its role as a host, it can be assumed that the media coverage of the Durban climate talks had an influence on public discourse, bringing greater awareness of climate change in South Africa and on the African content.

Previous research has shown that the mass media is the key source of information and the main factor in shaping people's awareness of climate change.[7] Anabela Carvalho and Jacquelin Burgess[8] have also underscored the role of the media in shaping public risk perceptions and social action on climate change. Given the immense importance of the issues at stake for people and the planet, this is a role that carries great responsibilities,[9] especially in Africa, where 'the media's role in contributing to public understanding of climate change is crucial'.[10] However, the editor of a recent volume on climate coverage in South African media 'paints a rather depressing picture of journalism's neglect and failure in the face of one of the biggest challenges of our time'.[11] One reason is that environmental issues often require further investigation and more external input than many other subjects, and because its impacts are less immediate and direct than many other pressing concerns, such as unemployment and crime.[12]

The historical responsibility for the carbon emissions that drive climate change rests on rich industrialised nations, which have benefitted from high levels of emissions from their own development. Today's high-income countries are responsible for roughly two-thirds of the emissions emitted since the Industrial Revolution, while counting for only about a sixth of the world's population.[13] Even though the greater part of the growth in emissions since 1970 comes from developing countries, the overall levels of greenhouse gases (GHG) remain larger in the rich industrialised countries. The average person from the North accounts for more than four times the carbon dioxide emissions of an average person from the South due to more energy-intensive activities.[14] Furthermore, poor people and poor countries, which have contributed the least to the problem, are hardest hit by climate change and will have more difficulties in adapting to future changes than rich people and rich countries. Africa, the Middle East, and South Asia are among the areas most vulnerable to climate change, due to drought, extreme weather, and sea level rise.[15]

---

7   A Carvalho 'Media(ted) Discourses and Climate Change: a Focus on Political Subjectivity and (Dis)engagement' (2010) 1 *WIREs Climate Change* 172.
8   A Carvalho & J Burgess 'Cultural Circuits of Climate Change in UK Broadsheet Newspapers, 1985–2003' (2005) 25 *Risk Analysis* 1457.
9   HDR (note 1 above).
10 H Wasserman 'The Challenge of Climate Change for Journalism in Africa' (2012) 33 *Ecquid Novi: African Journalism Studies* 1.
11 Ibid 2.
12 Schreiner & Bosman (note 6 above) 70.
13 WDR (note 1 above) 3.
14 For a more detailed overview see <http://hdr.undp.org/en/reports/global/hdr2011/summary/ trends/>.
15 Intergovernmental Panel on Climate Change (IPCC) *Fourth Assessment Report* Part 1 (2007).

The inequity in the global distribution of emissions and the current and future damage caused by climate change raises questions about climate justice, responsibility, and human rights and provides a backdrop for a global climate justice movement to emerge. The movement aims to raise awareness about justice issues among the public, to mobilise people from both the South and the North, and to demand climate justice in international climate negotiations. Climate justice demands that the industrialised North provide compensation to developing countries for the cost of reducing vulnerability and for the damage caused by climate change.

South Africa is in an unusual position where it straddles the 'carbon divide' between industrial and developing economies. On the one hand, South Africa is a relatively developed, economically powerful country, ranked as the 12th largest GHG emitter in the world.[16] The high fossil fuel intensity in South Africa is due to coal-fired power plants, which constitute the country's major energy source and supply the mining sector with cheap electricity. The mining sector is key to the country's economy, providing employment and export revenues. On the other hand, South Africa is a developing country with domestic challenges that run deep, which means it needs economic growth to reduce poverty. Forty-three per cent of the population lives on less than US$2 a day, and many will therefore argue that the first priority should be lifting millions of South Africans out of poverty. According to the official labour market statistics, 30 per cent of South Africans are unemployed. Hence, the country's government needs to strike a balance when mitigating climate change and simultaneously creating jobs, reducing poverty, and closing the income gap between rich and poor. This is echoed in South Africa's position in the international climate negotiations.[17]

This chapter addresses how a selection of South African media covered COP17 with a special focus on representations of climate change and links to poverty, justice, and human rights issues. In particular, I will address: How did South African media construct representations of climate change and interrelated development issues during the Durban climate talks? Which groups of social actors were likely to be portrayed in the news discourse as agents and as such most likely to hold the 'definitional power' of climate change in the public debate? How were the affected parties portrayed by the news media? To what extent, and how, was climate change framed as a climate justice issue and even a human rights issue related to the poor? I will explore these questions by using a discourse analytical approach, with both qualitative and quantitative elements. I will present an overview of salient points regarding climate change and poverty, followed by a discussion of climate ethics, justice claims, and human rights as a way to understand the context within which climate change is linked to poverty and justice issues. First, I will briefly

---

16   HDR (note 1 above).
17   Nationally appropriate mitigation actions (NAMAs) to be implemented by developing countries under UNFCCC <http://unfccc.int/documentation/documents/advanced_search/items/6911.php?priref=600006178#beg>.

outline how the media landscape has changed since apartheid, which also has a bearing on the media's coverage of climate change.

II THE SOUTH AFRICAN MEDIA LANDSCAPE AFTER APARTHEID

The media landscape has undergone fundamental changes since the end of apartheid. Essentially, authoritarian control of the media for political means has given way to press freedom,[18] globalisation, and marketization.[19] Herman Wasserman describes how the South African media had to reorient itself to a new global landscape, including penetration of global capital, but it also had to redefine its role domestically, in relation to a contested public sphere marked by severe class inequalities and competing normative expectations of what the media's role in the new democracy should be.[20] Even though a liberal democratic perspective of 'watchdog' journalism seems to dominate the commercial media landscape, a 'free press' is under pressure from the African National Congress (ANC) government that wants to establish a statutory Media Appeals Tribunal as an alternative to the appeal system in the current self-regulatory process.[21]

In a South African context, a strict public/commercial distinction between media outlets is not always straightforward. For example, the South African Broadcasting Corporation (SABC) has a mandate to provide public service broadcasting, but since it receives very little financial support from the government and license fees, the broadcaster relies heavily on advertisements, which incorporates commercial programming and journalistic orientations.[22] The SABC has been accused of having close ties to the ruling party, but concrete cases have been critically exposed and countered by voices in other media.[23]

The country's current media system is characterised by diversity, with more than 20 daily newspapers and hundreds of community newspapers. Households with televisions have increased significantly and now surpass households with radios.[24] There are three main players in the television (TV) market, including the SABC with three channels (*SABC1*, *SABC2* and *SABC3*).[25] In addition to the state-owned SABC, the media market is dominated by a handful of large corporations (Media24, Independent News and Media, Avusa Media, and the Caxton and CTP Group), which together own almost all the major and community newspapers. While the traditional quality newspapers remain

---

18 The South African Constitution, adopted in 1996, explicitly protects the freedom of the press, the freedom of expression, and access to official information in its Bill of Rights.

19 H Wasserman 'China in South Africa: Media Responses to a Developing Relationship' (2012) 3 *Chinese J of Communication* 336, 340.

20 H Wasserman 'Globalized Values and Postcolonial Responses: South African Perspectives on Normative Media Ethics' (2006) 68 *The International Communication Gazette* 71.

21 Wasserman (note 19 above) 340.

22 Ibid 341.

23 KS Orgeret 'South Africa: A Balancing Act in a Country of (at least) Two Nations' in E Eide, R Kunelius & V Kumpu (eds) *Global Climate – Local Journalisms* (2010) 291, 292.

24 South Africa Census 2011 <http://www.statssa.gov.za/Census2011/Products.asp>.

25 Read more <http://www.mediaclubsouthafrica.com/index.php?option=com_content&view=art icle&id=110:the-media-in-south-africa&catid=36:media_bg%20#ixzz2C1S3JL7e>.

elite-oriented with a limited circulation, a newly emergent tabloid press has been very successful in penetrating their target groups – the black working class.

However, serious coverage of climate change in the tabloid press is very limited. A previous study of the South African press coverage of COP15 revealed that the tabloid and top-selling daily newspaper, *The Daily Sun,* only published seven articles related to the summit or climate change, which were mainly limited to pictures and short paragraphs.[26] In contrast, *Business Day* had extensive coverage of COP15 (77 articles) and had reporters on the spot in Copenhagen.[27] A recent study found the same discouraging result from an investigation of *Son's* (*The Daily Sun's* Afrikaans cousin) COP17 coverage. Even though COP17 took place on South African soil there were only five articles about the climate summit.[28] The scanty coverage in the popular press suggests that climate change is perceived as an elite issue, which is rather ironic since it is precisely the person in the street who will be hardest hit by climate change.

III   CLIMATE CHANGE AND POVERTY

The Millennium Development Goals (MDGs) aim to eradicate extreme poverty and hunger, improve health care and education, promote gender equality, and ensure environmental sustainability. Though much has been achieved over the past decades, climate change now hampers efforts to deliver on the MDG promise. Looking to the future, the danger is that climate change will stall and then reverse progress that has been made over generations – not just in cutting extreme poverty, but also in health, nutrition, education, and other areas.[29] South Africa has made significant progress over the past 15 years on several living standards indicators, such as the increase of households living in formal dwellings, access to piped water, and the use of electricity for lighting, as well as progress in education.[30] However, South Africa's past of racial inequality is still clearly evident, which adds to the urgency of addressing socio-economic inequalities. In this context, climate change may seem distant compared to more urgent problems, but if climate change is not addressed properly it will just reinforce poverty and socio-economic inequalities.

Across developing countries, millions of the world's poorest people are already being forced to cope with the impacts of climate change. In Africa, loss of rainfall and changing rain patterns, desertification, and drought are creating problems in an already stressed environment. Recent warming appears to have enhanced drying over many land areas during the past decades.[31] The lack of irrigation possibilities threatens the production of food

---

26   Orgeret (note 23 above) 294.
27   Ibid.
28   Unpublished paper presented by Alet Janse van Rensburg at IAMCR conference in Durban 2012.
29   HDR (note 1 above) 1; see also UNDP 'Human Development Report (HDR) 2011: Sustainability and Equity: A Better Future for All' (2011).
30   See Census 2011 (note 24 above).
31   A Dai 'Increasing Drought under Global Warming in Observations and Models' *Nature Climate Change* (2012).

and supplies of water and energy. Crops in certain African countries may be halved in the course of only a few decades.[32] Increased temperatures may also lead to more heat-stress related deaths in livestock. South Africa is expected to get 20 per cent less rainfall,[33] which along with reduced soil moisture will lead to the danger of extreme droughts in this century.[34] Changes in rainfall patterns and river flow, more flooding and drought will make the livelihoods of poor people more precarious, and infectious diseases are projected to rise along with malnutrition and hunger. Adverse environmental factors are expected to boost world food prices 30 to 50 per cent in real terms in the coming decades and to increase price volatility, with harsh repercussions for poor households.[35]

South Africa and the rest of the African continent face particular risks due to their economic dependence on climate-sensitive sectors such as agriculture, forestry, and fishing (ocean warming will affect coastal marine fisheries), all of which employ a relatively high number of people when compared to their outputs. Storms, coastal erosion, and sea level rise may damage coastal zones, including large cities (particularly Cape Town and Durban) with impacts on settlements, transport systems, energy, and industry.[36] As elsewhere, urban poor in squatter settlements, as well as rural poor who depend on natural resources for their livelihoods, are at particular risk for weather-related disasters, which in turn will make their poverty deeper and harder to escape. To the extent that women are disproportionately involved in subsistence farming and water collection, they face greater adverse consequences of environmental degradation.

IV   CLIMATE ETHICS, JUSTICE CLAIMS AND HUMAN RIGHTS

Climate justice is used as a term for framing climate change as an ethical issue and considering how its causes and effects relate to concepts of justice, equity, and human rights.[37] Climate justice calls for the recognition of a principle of ecological debt that industrialised governments and transnational corporations owe the rest of the world as a result of their appropriation of the planet's capacity to absorb GHGs. The debt carries an obligation to protect the rights of victims of climate change and associated injustices to receive full compensation, restoration, and reparation for loss of land, livelihood, and other damages.[38]

---

32   IPCC (note 15 above).
33   Ibid.
34   Dai (note 31 above).
35   See <http://hdr.undp.org/en/reports/global/hdr2011/summary/>.
36   MD Smith *Just One Planet. Poverty, Justice and Climate Change* (2006) Appendix A.
37   Climate justice is also used with reference to legal systems, where justice is achieved through the application and development of law in the area of climate change <http://www.climatelaw.org/>. However, in this chapter it is not used as a legal term. See also <http://en.wikipedia.org/wiki/Climate_justice>.
38   Bali Principles of Climate Justice (29 August 2002) <http://www.indiaresource.org/issues/energycc/2003/baliprinciples.html>.

Initially, climate justice was a concept held by religious organisations, but since the first Climate Justice Summit took place in The Hague in 2000, parallel to the Sixth Conference of the Parties, it has been embraced by a much wider group of activist organisations, networks and grassroots initiatives working for social and/or environmental justice. In particular, a climate justice movement grew and became more visible prior to COP15 as an intriguing mix of coalitions seeking to mobilise governments and the public to tackle climate change. In the swarm of messages, the attempt to redefine climate change from a purely scientific issue to a human rights and environmental justice issue sticks out. The movement is trying to put a human face on climate change and make links to local communities.[39]

Prior to COP17, international and local groups prepared to make the climate justice movement visible. During the conference, different groups engaged in street demonstrations, educational side events, and campaigns. For example, Climate Justice Now! is a network of 1,000 organisations which demands that industrialised nations implement drastic emission reductions, increase financing to support adaptation programmes in the developing world, and support rights-based conservation programmes that promote community control over energy, forests, and water. It can be argued that the movement's overarching principles, which call for climate equity, inclusive participation, and human rights have yet to be a defining factor in the arena of global climate change policy. Nevertheless, several issues pushed by the movement's organisations have been incorporated in the United Nations Framework Convention on Climate Change's (UNFCCC) framework, including the need for adaptation funds, disaster risk reduction, and the need to integrate the concerns of marginalised groups such as indigenous people into its programmes.[40]

In a world increasingly threatened by climate change it is becoming obvious that its impacts undermine, and are probably already limiting the attainment of a wide range of universally accepted human rights: rights to life, health, food, water, shelter and property, rights associated with livelihood and culture, as well as implications of climate-induced displacement and conflict.[41] People living in extreme poverty are expected to be the hardest hit by climate change, but these are also the people whose rights are already precarious. Although South Africa has a progressive Constitution with its own Bill of Rights[42] the same situation is true there. In particular, black urban shack dwellers are at risk as are the rural poor who depend on natural resources for their livelihoods.

However, adopting a human rights framework in international climate change policies presents many complexities. Climate change is distinguished from other environmental hazards due to its global nature and temporal

39   See <http://www.indiaresource.org/issues/energycc/2003/humanfacehumanproblem.html>.
40   See <http://www.climate.org/climatelab/Climate_Justice_Movements>.
41   See, for example, the Report of the Office of the United Nations High Commissioner for Human Rights on the relationship between climate change and human rights <www.ohchr.org/EN/Issues/HRAndClimateChange/Pages/Study.aspx>.
42   See <http://www.info.gov.za/documents/constitution/1996/96cons2.htm>.

dispersion, which blur both the causes and the effects of climate change.[43] It is therefore not possible to directly link one perpetrator to one specific victim; instead, tiny footprints are left all over because the atmospheric concentration of CO_2 does not enable us to distinguish the source of the emissions. It is therefore easier to use human rights in a concrete case where the relationship between the culprit and the victim are direct and perceptible, such as in the case of an oil spill. As Stephen Humphreys argues, human rights law does not easily reach across international borders to impose obligations in the case of climate change.[44] Broad human rights mean that the state is obligated to protect its citizens. In the international climate negotiations under UNFCCC, states' obligations are primarily toward one another. In line with this, I will argue that climate justice demands related to human rights, especially in the context of an international climate regime, are rooted in an ethical judgement, which provides for authoritative advocacy. In other words, a climate justice movement incorporates human rights rhetorically to mobilise support as part of a broader approach of confronting major polluters.

(a) Research questions

The mass media plays an important role in making the world aware of the human consequences of climate change and the need for protection from the accelerating and damaging consequences of climate change, which will make it harder to fulfil human rights obligations to poor and vulnerable people. An issue at stake is how media representations at the nexus of climate change, justice, and right views on poverty might encourage or limit the possibilities of interpretations. This raises a series of research questions, which can be summarised as follows: (1) How did South African media, linked to different political and economic positions, construct representations of climate change and interrelated development issues during the Durban climate talks? (2) Which groups of social actors, representing different interests, were likely to be portrayed in the news discourse as agents and as such most likely to hold the 'definitional power' of climate change in the public debate? (3) How were the affected parties portrayed by the news media? (4) To what extent, and how, was climate change framed as a climate justice issue and even a human rights issue related to the poor?

Overall, the aim is to contribute to a better understanding of how climate change challenges are constructed and framed by the media as embedded in narratives and news conventions. This understanding can help reveal relationships of definitional power and their implications for pro-poor development.

V   A DISCOURSE ANALYTICAL APPROACH

I have chosen a discourse analytical approach in order to understand how South African media constructed the meaning of climate change and related

---

43   J Garvey *The Ethics of Climate Change. Right and Wrong in a Warming World* (2008) 59.
44   S Humpreys (ed) *Human Rights and Climate Change* (2010) 5.

development issues during the Durban climate talks. Discourse analysis can loosely be defined as a systematic and explicit study of the structures and social or cultural functions of media messages, understood as specific types of text and talk.[45] Scholars working with this kind of analysis suggest that stories and arguments draw on a relatively fixed repertoire of linguistic strategies, combining premises and conclusions, assertions and substantiation, scenes, actors, and themes.[46] Discourse analysis shares with framing analysis an interest in the social construction of a phenomenon (for example climate change) by mass media, recognising that journalists do not only convey 'objective' news stories, but also establish interpretative schemes in communication with their readers within which those stories acquire meaning. Even though this chapter does not pursue a framing analysis, the concept offers a useful way to describe the power of communicating a text. As defined by Robert Entman:

> to frame is to select some aspects of a perceived reality and make them more salient ... in such a way as to promote a particular problem definition, causal interpretation, moral evaluation, and/or treatment recommendation for the item described.[47]

Frames can limit debate by establishing the vocabulary and metaphors through which participants comprehend and discuss an issue. In particular, I find the concept of framing beneficial in examining and discussing to what extent climate change was 'framed' as a climate justice issue or in broader terms, within a morality frame versus a technological and economic frame.

In developing an *analytical model* for analysing media texts, I draw upon elements from narrative analysis combined with lexical examination. Narrative analysis provides an alternative in qualitative analysis, which has been haunted by the problems inherent in addressing the quantity and heterogeneity of media texts due to the limited number of texts. In narrative analysis, large groups of texts can be broken down into their basic components and structures by means of qualitative procedures – without breaking up the text as a meaningful whole.[48] News is not a story in a strict narrative sense, but journalists are storytellers that need events, characters, and settings. Rather than being based on a logical chain of events, most news stories describe a single event without elaboration about causal connections. A hierarchical set of topics forms the thematic structure of the text.[49] At the top of this macro-structure we find the headline and lead paragraph, which summarises the text and specifies its most important information, followed by main events, context or history, verbal reactions by informants and possibly the journalist's

45 TA van Dijk 'The Interdisciplinary Study of News as Discourse' in KL Jensen & NW Jankowski (eds) *A Handbook of Qualitative Methodologies for Mass Communication Research* (1991) 108. Discourse analysis entails a variety of theoretical entrances and methods. See, for instance, in addition to Van Dijk, N Fairclough *Media Discourse* (1995) or R Wodak & M Meyer 'Critical Discourse Analysis: History, Agenda, Theory and Methodology' in R Wodak & M Meyer (eds) *Methods of Critical Discourse Analysis* (2001) for input with regard to different approaches.
46 M Coulthard & M Montgomery (eds) *Studies in Discourse Analysis* (1981).
47 R Entman 'Framing: Towards Clarification of a Fractured Paradigm' in D McQuail (ed) *McQuail's Reader in Mass Communication Theory* (2003).
48 S Chatman *Story and Discourse: Narrative Structure in Fiction and Film* (1978).
49 Van Dijk (note 45 above) 113.

comments. The relevance structure might favour attention to some aspects, while leaving out other information and evaluations about a situation, event or issue. The analysis can mainly be divided into four parts.

(1) *Main topic*: I have recorded the main topic and sub-topic for each text, along with date, page/section/running time, etc. I will primarily be concerned with the main topic, as this comprises the most salient information in the news item, typically contained in the headline and lead paragraph. Topics were divided into eight categories, which made it possible to look at the quantitative distribution of the topics across different outlets.

(2) *Thematic structure* and *form:* The thematic structure or story of a text consists of an event or chain of events, behind which one or more *agents* (news actors) stand.[50] An agent is the narrative subject of the narrative predicate, described in statements in the form of 'do'. The importance of an agent can be seen to the extent that he or she takes significant action. Major agents are those who play a leading part in the stories, while minor agents could have been omitted without destroying the story even though it would impoverish it. I distinguish between major and minor agents, but the former will receive precedence. Each text included in this research has been systematically subjected to an analytical schema, enabling me to tally the story elements: events (what happens to whom), agents and existents (characters and setting). This approach makes it possible to extract information across the news stories. In a similar manner as for topics, I have categorised agents into groups, which shows the similarities and differences between media outlets. This in turn may be linked to their distinct political and economic positions.

(3) *Identifying climate justice claims:* I find Humphreys' outline[51] of four justice claims useful to identify and distinguish between different aspects of climate justice in concrete media texts, particularly because they are relevant to the international climate change negotiations.[52] I have added ecological justice to the list and also put more stress on the intergenerational aspect than Humphreys. The following claims will be identified and tallied together with who is promoting them: (a) *Corrective justice* arises due to the activities of one group of people – those who overuse the global carbon dump[53] – which have caused and continue to cause injuries that affect a different (and larger) group of people who live in parts of the world likely to be hardest hit by climate change. Possible solution: international funding (for example Green Climate Fund, Adaptation Fund); (b) *Distributive justice and procedural justice*

---

50   Chatman (note 48 above).
51   Humpreys (note 44 above).
52   I don't have the space here to go into details about the claims, please see Humphreys (note 44 above) for more information.
53   The term 'global carbon dump' captures the notion that the atmosphere can support only a limited amount of GHGs – and so there can be no unrestricted right to send carbon into it (Humphreys (note 44 above) 24).

involves the global allocation of costs and benefits and the burden-sharing of mitigation *and* mechanisms that will ensure that a just solution can be reached, respectively; (c) *Substantive and intergenerational justice* is associated with the need for economic growth and concern for future opportunities. A global freeze or steep reductions in the use of fossil fuels may lock-in vast wealth disparities between groups in different regions. Possible solution: financial support and technology transfer to provide low-carbon intensive growth (for example the Green Climate Fund); (d) *Entitlements* are derived from prior usage, that is, 'legitimate expectations'. The choice of fossil fuel to power the Industrial Revolution was not made with an understanding of its subsequent outcomes, nor was anyone able to foresee these outcomes. This means that some countries, in particular those that depend on fossil fuel exports, may use versions of this argument to claim that they deserve entitlements in the distribution of burdens;[54] (e) *Ecological justice* can be defined as justice for animals, living beings and living ecological systems as well as humans.[55] I will define this claim as non-human justice, because humans are covered in the other claims. However, in reality they will often overlap and intervene.

(4) *Lexical analysis:* I have chosen to complement the findings with a lexical investigation to expose choices of vocabulary in the media texts, which may convey and contribute to a world-view, or a pattern of attention and perception.[56] Some may argue that scanning for certain key words or phrases has limited value, because similar concerns may be addressed using different terms, this approach determines to some extent the relevance of certain kinds of information, orientations, and responses. For example, scanning for 'human rights language' says something about the salience of viewing climate change as a human rights issue.

All in all, the analytical framework should help to illustrate how climate change is defined as a problem, who is seen as the important actors and what or who is considered worth protecting (for example nature, future generations or today's poor), and finally what a possible solution might look like. I will deliberately check for textual interrelationships to ensure completeness.[57] In addition to the textual level, I relate findings to an extra-textual framework for explanation, which together should make for solid conclusions.

(a) Choice of outlets

The media outlets in this study are the SABC's 'flagship', the *SABC3* evening TV news, *Business Day*, and *The Mercury*. I have chosen the SABC because it has a mandate to provide public service broadcasting and has the widest audience due to its programming diversity. As a host broadcaster, its news headquarters was based at the COP17 venue in order to serve around a

54 Garvey (note 43 above) 59.
55 NP Low & BJ Gleeson *Justice, Society and Nature: An Exploration of Political Ecology* (1998).
56 RJ Alexander *Framing Discourse on the Environment: A Critical Discourse Approach* (2009).
57 Wodak & Meyer (note 45 above) 31.

thousand international media representatives with television feeds as well as facilitating live coverage in South Africa.

Furthermore, the economic implications of the outcome of the international climate negotiations are obvious, whether there is a lack of action to curb emissions or restrictions that force countries to downscale or alter their economic endeavours. Hence, I have chosen a prominent national daily business newspaper, *Business Day*, which reports on business subjects, politics, and current affairs, including energy, agriculture, health, and environmental news. It does not derive its influence from its reader numbers, but by targeting the country's powerful elites and decision-makers. It is owned by Avusa Media, which is known for strong online presence. According to the owner's homepage, all its media titles seek to uphold the values of the Constitution and are guided by a code of conduct that is closely aligned to the South African Press Code.[58]

Finally, the Durban morning newspaper, *The Mercury*, provides regional level coverage. It is a traditional quality newspaper, and is clearly chosen for its location, because COP17 took place in Durban. It is targeted at middle- to upper-income readers. The most recent figures from the All Media Products Survey (AMPS) suggest a rather even distribution between black, white, and Indian readers (numbers received from the newspaper). *The Mercury* is owned by the Irish-based Independent News and Media PLC, which controls 14 national and regional newspapers in South Africa. It has no overt political leanings, although readers from diverse political persuasions might disagree with that characterisation for different reasons.

I have chosen media outlets that can be expected to provide solid coverage of the climate talks and thereby play an important role in defining climate change in the public debate. Hence, I have omitted any of the tabloid newspapers given their poor record in reporting on serious issues such as climate change. Moreover, the selected outlets provide a contrast in terms of political and economic orientation, and editorial and linguistic profiles. They also serve different audiences. We can probably expect that *Business Day* will tend to see the international climate negotiations in economic terms, as this would be an interest of their affluent audience, while the *SABC3* news will probably be closer to the official political agenda and more people-oriented. *The Mercury* team was led by the award-winning environmental writer Tony Carnie, known for his fearless and thorough reporting and refusal to step away from controversy, which raises the expectations of critical reporting promoting climate justice issues. All three media use English, which is often considered the country's lingua franca because it is the second language of most South Africans; hence, many discussions of local and national importance take place in English language media.[59]

I have included press coverage from a few days before the climate summit to survey some of the expectations and background material, as well as during the summit, and from a few days after the summit, which typically provide

---

58   See <http://www.presscouncil.org.za/ContentPage?code=PRESSCODE>.
59   Orgeret (note 23 above) 292.

the first interpretations of the negotiations (25 November 2011 to 14 December 2011). Both printed and online articles from the two newspapers in my study are included in this investigation (but not duplicate articles). With regard to *SABC3* evening news, the period covered is from 25 November 2011 to 12 December 2011, as well as a news broadcast on 2 October 2011, which covered the last official UNFCCC meeting in Panama in preparation for COP17, as well as 2 November 2011, concurring with a pre-COP17 summit, which clearly stated South Africa's aspirations one week ahead of COP17.[60]

VI ANALYSIS

The media outlets in this study put extensive resources into covering COP17, which is reflected in their output. *SABC3* placed the climate talks high on the news agenda during the whole period of investigation, with a total of 18 news items. The broadcast news stories were long, most lasting around five minutes and containing two or three storylines (total 44 story lines). *The Mercury* also made COP17 its chief concern. In addition to its regular news product, the newspaper published a daily COP17 pullout supplement with softer background stories and recapping. In total, 160 articles were published in print or online 20 days before, during, and after the event (duplicates and images were not counted). In order to capture the editorial concerns of *The Mercury* I have included 111 articles for further scrutiny, excluding non-staff materials and readers' letters. In comparison to *The Mercury* and *SABC3* news, *Business Day* has a history of covering the international climate negotiations. During the COP17 period, the newspaper published 110 articles, compared to 77 articles during COP15. I have included 101 articles, omitting readers' letters.

In total, 211 newspaper articles and 18 news broadcasts have been analysed. On average the newspapers each published five articles per day. The coverage was dominated by the news genre, but there were also a significant number of opinion articles written by staff in both newspapers. The newspapers had at least one editorial each and commentary/analysis comprised roughly one of ten articles.

The prominence given the climate talks was also reflected in the headlines. Most of the news broadcasts made it to the headlines and COP17 was the number one headline on at least six occasions during the event. *The Mercury* put seven stories about the issue on the front page. In *Business Day* the attention was more subdued with only two stories on the front page, which stands in stark contrast to the newspaper's priority during COP15, when 21 stories were given space on the first page.[61] It is conceivable that this downscaling is part of an international trend where attention to climate issues has diminished since the failure of the Copenhagen talks. In the following, I will present the

---

60 Unfortunately, the recordings of the SABC news on 2 and 11 December 2011 were corrupted and had to be removed from the sample.
61 Orgeret (note 23 above) 295.

findings of the media's prioritisation of topics, followed by a discussion of social actors, justice claims, and lexical choices.

(a)   Main topic: more to COP17 than the political game

The distribution of topics, as shown in Figure 1, reflects the fact that COP17 was primarily a major political event. It was therefore natural that political solutions, the political game, or positions taken by the different parties topped the list of topics reported. The prominence of politics was also reflected in that most of the stories that made it to the front page or as TV headlines were concerned with the political side of the climate negotiations.

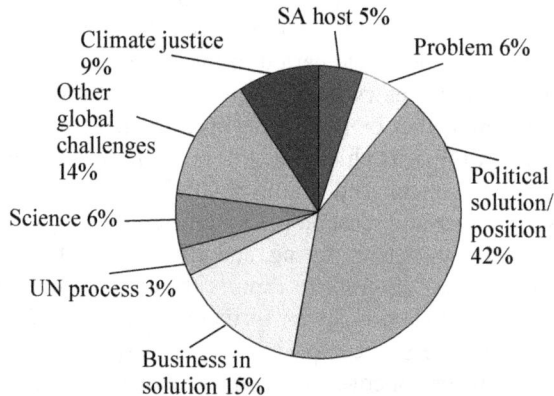

Figure 1: Media coverage by main topic, all outlets (N=229)

However, the media outlets also gave ample space to stories that had different approaches to climate change. Roughly half of the articles did not focus directly on the negotiations, but on side events, demonstrations, or other related stories. It is notable that the business role in the solution represented a big piece of the coverage, but also that climate justice and other global or development challenges linked to climate change were accounted for.

The majority of the stories on the business role in solving climate change related to green solutions, renewable energy, and innovations. Both private and government stakeholders were seen as important in creating investments in a low-carbon economy. A related theme was the creation of green jobs and balancing mitigation strategies with securing employment. Articles might acknowledge challenges in the South African economy, as illustrated by: 'South Africa should industrialize renewable energy sector',[62] which is linked to the coal-dependent economy, but still forward-looking. A similar view was stated by President Zuma:

> This renewable energy project also confirms ... our view that we cannot separate climate change responses from our goals of pursuing development and poverty eradication. Pursuing the green economy must be linked to our overall agenda of pursuing employment creating growth.[63]

---

62   *Business Day* (2 December 2011).
63   *SABC3* (4 December 2011).

Climate justice and other global challenges together constituted the main topic in one of four stories. Other global challenges are a diverse group of issues linked to the intersection between climate change and poverty or development. One concern seemed to be especially worrying across all media outlets, which was the impact of climate change on food security, including crop failure, rising food prices and hunger. Agriculture was presented as an important part of the solution. Vulnerability, adaptation, and improving resilience were also recurring issues, and addressed subjects across a range of different sectors.

The political and commercial orientations of the different media outlets might lead us to expect the SABC's coverage to be closer to the official political agenda, while the mainstream print media's professional ideology of being a watchdog and 'unofficial opposition' to the ANC government would lead them to be sensitive to claims from civil society. Figure 2 shows that this was only partly true.

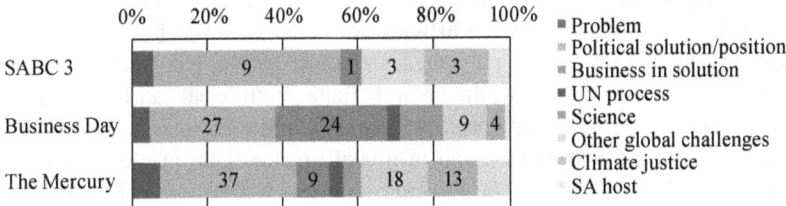

Figure 2: Distribution of main topics by media outlet (N=229)

As presumed, the majority of the broadcasts gave first priority to political issues. However, there were also a considerable number of broadcasts that had climate justice and other global challenges as the main narratives (almost one of three). This was also reflected in the headlines in which activists' demands or protests were salient: 'Thousands demand action on climate change as the streets of Durban reverberate with protest'; 'Tempers flare at COP17 as activists and police crash'; '... activists turn up the volume as COP negotiators burn the midnight oil'.

These headlines suggest that the SABC was not only following the political agenda, but was also sensitive to activists' concerns. Yet I suspect that part of the explanation is due to television's need for live images, preferably lively images that can create momentum. Obviously, thousands of people in the streets demonstrating or doing different stunts is more eye-catching than grey suits. The vigorous jargon in the headlines (my emphasis) builds on this action-oriented frame. My suspicion was strengthened by a conversation I had with a central SABC-TV news reporter at the end of COP17, which can be illustrated by the following statement:

> You remember the Greenpeace people that were deported after their acts at the hotel where the business leaders were holding a meeting? Civil society really generates a lot of great stories for us in the media, they give us a lot of action.

The finding also conforms to the previous discussion in which the SABC had to 'lean towards commercial orientations'.

Business Day was the least concerned with the political agenda and climate justice and gave priority to viewing climate change through an economic lens, which is in keeping with the presumption that this would be of interest to their affluent audience. Nonetheless, there was still some room for other global/ development challenges and to a more limited extent climate justice as a core theme, which I will return to later. *The Mercury* lived up to its watchdog role.

It had three times as many articles concerned with climate justice and twice as many articles about causes triggering climate change than *Business Day*, which were topics that often take a more critical viewpoint of a situation. *The Mercury* reported on the 'bad guys': Eskom and Sasol and coal-fired power stations in general, big corporations, and banks that lent money to fossil fuel energy and mining, as in 'Fossils are found wanting, say activists'.[64]

Across the three outlets it was South Africa's reliance on coal-powered electricity that worried the media the most, but the implications of a South African carbon tax on jobs, Canada's tar sand projects, and deforestation were also reported. A handful of the articles presented criticism towards market mechanisms. *The Mercury* published an article concerned with corporate greed and market fixes of the World Bank with clear anti-capitalistic undertones 'Activists *decry* corporate *greed*',[65] while *Business Day* warned against the consequences of a gap in global finance 'Gap in global climate finance will be a disaster'.[66] *The Mercury* reporters wrote more pointedly than their peers at *Business Day*, which can be seen in the selection of stories, as well as how they angled their stories, in their choice of sources, and their more emotionally charged language (see my emphasis). For example, both newspapers published a critical story on Reducing Emissions from Deforestation and Forest Degradation (REDD), see excerpts from *Business Day* and *The Mercury*, respectively:

Bolivia's forest proposal gets little attention
Bolivia has tabled a new proposal on forest conservation, but has so far been disappointed by the lack of attention from other negotiators ... Bolivia's chief negotiator, Rene Orellana, said his country was *worried* about proposals to use private sector markets to fund forest conservation, as this did not take account of the multiple functions forests provide.[67]

Forests are not for carbon stocks
Bolivia came out *swinging* at its first press conference ... *opposing* REDD. Bolivia is showing *strongly* against the mechanism of REDD ... 'As people who live in the forest, we are not *carbon stocks*. We disagree with REDD because we oppose the *commodification* of the forests', said Orellana.[68]

Business Day confined itself to translating the negotiator's statement as a note of concern; the negotiator was *worried* about using private sector markets to fund forest conservation. *The Mercury* story presented stronger criticism of market-based approaches, which can be seen in the headline and main entry as well as the language style (see my emphasis). This impression is further

---

64   *The Mercury* (7 December 2011).
65   *The Mercury* (2 December 2011).
66   *Business Day* (2 December 2011).
67   Ibid.
68   *The Mercury* (note 65 above).

strengthened in the following article: 'Forest plan just a "giant land grab"'[69] in which Berenice Sanchez, from Mexico, argued that extractive industries would be rewarded with carbon offsets for 'raping mother Earth'. The article uses very negative language, such as 'rape', 'giant land grab' and 'new form of colonialism'.

The Mercury clearly prioritised *climate justice* issues, which counted for 12 per cent of the newspaper's COP17 coverage. It embraced events that were organised by the climate justice movement, such as the Trans African Climate Caravan of Hope;[70] the Youth Caravan: We have Faith – Act Now for Climate Justice; and the inter-faith rally at Durban's King Park. Other subjects covered were protection of ecosystems, ecological credit, youth delegates protesting against Canada, and finally more general demands for a fair, ambitious, and binding climate deal. In contrast, *Business Day* covered a much more limited scope of issues, including civil opposition to the Canadian government and countering the global division in monetary terms, as in 'The North owes the South trillions in damages'.[71] *SABC3* news only scratched the surface, presenting climate activists' political demands for a fair deal, the voice of island states, and placards with 'Don't Kill Africa'.

The prominence given to climate justice and activists' claims in *The Mercury* was also reflected in two stories that made it to the front page: 'Only a revolution will do'[72] and 'Climate change activists arrested, deported'.[73] Both articles foreground the activists' point of view. The first article starts with 'Humanity has nowhere to go. "This is the only home. If you *destroy* it, it's finished"', which is a quotation from Archbishop Desmond Tutu, urging the nations of the world to join hands in Durban and find a way to prevent humanity's common home from being turned into a desert. After a brief introduction, the second article states 'Greenpeace Africa campaign director Olivia Langhoff said the activist had tried to climb Durban's Protea Hotel to hoist a banner reading "Listen to the people. Not the polluters", yesterday morning', a message that was expanded upon in subsequent paragraphs.

The Mercury had twice as many articles reporting on *other global or development challenges* than *Business Day*, but the latter also picked up on a range of development issues linked to climate change. However, there seems to be a difference in approach. Whereas *The Mercury* was more concerned with how climate change was affecting human development and to some extent the environment, *Business Day* gave more attention to how society could adapt to a changing climate. Some of the *Business Day* stories related adaptation to climate finance and the Green Climate Fund, while others illuminated the big challenges associated with climate change such as poverty, vulnerability, population growth, food security, agriculture, infrastructure

69   *The Mercury* (3 December 2011).
70   Organised to mobilise African civil society to articulate advocacy demands to the Africa people and its leaders and the world <http://www.pacja.org/index.php?option=com_ content&view=article&id=87:the-trans-african-caravan-of-hope&catid=7:home>.
71   *Business Day* (1 December 2011).
72   *The Mercury* (28 November 2011).
73   *The Mercury* (6 December 2011).

and land use, water resources and management, coastal management, waste management, and job creation. *The Mercury* stories presented climate change implications for environmental conservation, water security, poverty and loss of livelihoods, education, rural development, and women's empowerment. Climate-induced health effects, such as increased risk of waterborne illnesses and malnutrition, were reported equally in both newspapers.

(b)   Social actors: agents and existents

(i)   Agents – those who act

An important research question for this study is which groups of social actors, representing different interests, were likely to be portrayed in the media discourse as agents and as such most likely to hold the 'definitional power' of climate change in the public debate. To answer this question, I looked at the agents portrayed in each article. The key actors at international climate negotiations are top politicians and chief negotiators, who are also given the greatest power to define what climate change is all about in the mainstream media. As illustrated in Figure 3, they constituted the bulk of the major agents, and even more so in *SABC3* stories.

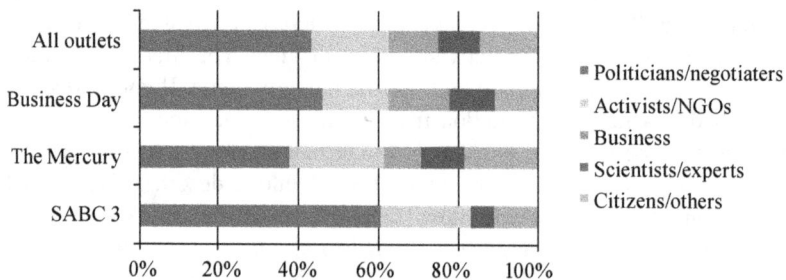

Figure 3: Distribution of major agent by media outlets (N=229)

However, bear in mind that even though COP17 was a political event where civil society did not have a formal voice, activists and different non-governmental organisations (NGOs) appear to have been successful in airing their concerns regarding the negotiations and related issues. Activists/NGOs comprised the second biggest group of agents in *SABC3* and *The Mercury*, while business agents were the second largest group in *Business Day*. As can be seen in the examples in Table 1, the choice of agents led to very different stories and frames for understanding climate change:

NGOs/activists and business representatives were positioned as active participants in the narratives of climate change. The first example shows a nuance in the title between the two newspapers, where *The Mercury* emphasised an NGO's role in the report, while *Business Day* chose to omit an agent (NGO) in the headline. Essentially, *Business Day* viewed the negotiations through an economic lens in which business/private sector has an important role to play in solving the problem – a much less threatening option than activists hammering on the door of the corporate world.

Table 1: Newspapers' choice of agents

The Mercury	Business Day
NGOs name and shame 93 top banks (2 December 2011)	Report names and shames 'climate killer' banks (2 December 2011)
Get moving faith leaders urge (28 November 2011)	Business must act now or lose out (29 November 2011)
Marchers call it like they see it (5 December 2011)	Business takes bull by the horns (5 December 2011)
Canadian youth marched out of COP17 (8 December 2011)	Business urged to take a lead on climate change (7 December 2011)
Kenyan activist lauded at COP17 (8 December 2011)	SA companies keep pace on green reporting (7 December 2011)
Activists decry corporate greed (2 December 2011)	Corporate SA at the forefront of a new revolution (9 December 2011)
People power our only hope (6 December 2011)	Car firms showcase electric future (30 November 2011)

However, it is important to underline that even though there are distinct differences between the two newspapers in the salience of topics and news actors and the language with which they were presented, the division is not watertight (for example, *Business Day's* 'Indigenous rights group turns up the heat on Canada'[74]). *Business Day* was also concerned with a range of global/development issues, but tended to see them on a more macro level. *The Mercury* had many articles on greening the economy, but the agent was sometimes omitted in the headlines, as in 'Out of necessity comes green innovation'.[75] However, there were also a few examples in *The Mercury* where private actors were portrayed as an agent for good or bad in the headline 'Shipping industry sees price on carbon emissions'[76] v 'Sasol tops list of emitters'.[77]

(ii) Existents – faceless victims

An existent is a social actor that is inscribed with a passive role in the narratives and as such is the opposite of an agent that plays an active part. Existents are typically those affected by climate change or other people's actions in which the effect can be positive or negative for the involved parties. A focus on the affected parties of climate change compels journalists to employ a human-interest narrative, which catches the interest of the readers/viewers. A human-interest frame usually entails an emotional point of view, such as how people have been personally affected by climate change. However, the journalists' search for drama (conflicts, alarming consequences etc) raises several challenges, because the potential effects of climate change have barely begun to unfold. It is a slow burning issue. There are also many uncertainties about what is caused by what, due to the contributions of natural and human-driven climate change. Hence, serious journalists will be careful

---

74 *Business Day* (7 December 2011).
75 *The Mercury* (7 December 2011).
76 *The Mercury* (29 November 2011).
77 *The Mercury* (note 75 above).

in attributing concrete events caused by extreme weather, drought or flood to climate change. This also leads to less substantiality in describing who is affected by climate change, which was also reflected in this material. Those affected by climate change were mainly referred to in general terms, and in this order, as Africa/Africans, poor people/countries, developing countries, small island states, or the least developing countries. Global terms were also used, but to a lesser extent. These terms included human civilisation, citizens of the globe, millions of people on the planet or future generations, including children and youths. The non-personalised, broader level of depicting the affected parties, such as cities, nations or continents in order to emphasise the dramatic outcomes of climate change is a general trend in climate change reporting.[78]

Quest for a better life in Africa

November 28 2011 at 11:17am

Leanne Jansen

leanne.jansen@inl.co.za

WITH her eyes closed and palm on her chest, Malawian farmer Agnes Raphael recalled the misery of having to abandon her education while still in primary school.

As evidenced by her swollen ankles, the 40-year-old travelled by bus for seven days to reach Durban in time for the 17th Conference of the Parties of the UN Framework Convention on Climate Change (COP17), which she hopes will ensure that her daughter does not also leave school early.

Raphael

Should her yield of Irish potatoes continue to dwindle – owing to what she believes to be climate change-induced drought – she will be unable to pay the 45 000 Malawian kwacha (R2 345) school fees each term.

As part of a Pan African Climate Justice Alliance initiative, Raphael, along with 200 other farmers, pastoralists, artists and activists from nine African countries, has made the trip to publicise the plight (and demands) of the continent to the international community.

Raphael, who also has four boys, said: "I plant crops, but there is no rain. There is not much to sell... If my girl can be educated, I will have been a good mother. I left school in the primary phase. I don't want any one of my children to leave school."

About a dozen of the news items were a bit more concrete and referred to vulnerable groups (besides the poor), such as rural communities and indigenous people, farmers and fishermen. Only *The Mercury* deployed a handful of stories in which people affected were of flesh and blood, most of whom were women. I have chosen to highlight the particular story of Agnes Raphael because it serves well to distinguish a *Mercury* story. It is about a Malawian farmer who travels with the 'Caravan of Hope' to tell COP delegates and the world how drought is affecting her yields, so she may not be able to pay school fees for her children (which implies that the daughter

---

78   A Dirikx & D Gelders 'To Frame is to Explain: a Deductive Frame-analysis of Dutch and French Climate Change Coverage during the Annual UN Conferences of the Parties' (2010) 19 *Public Understanding of Science* 732, 734.

will be the first to lose out). This story presents climate change as a risk for real people, which can hamper progress in education and girls' empowerment. And Raphael was not alone; she travelled with 200 other farmers, pastoralists, artists, and activists from many African countries who rallied together to demand for climate justice. This brings me to a second point. Even though Africa or African people were portrayed as victims of climate change in this and many other stories in this study, they are also presented as agents who fight for justice and their lives. This is also highlighted a little further down in the article in which the participants of the Trans African Climate Caravan of Hope were observed singing the refrain, 'Why, why, why? Africa is a victim. Now, now, now. Climate justice now'. In general, developing countries and Africa in particular were not portrayed as helpless as a consequence of their vulnerability; rather people were shown as fighting for justice and to develop their countries.

(c) Justice claims

Table 2 shows which justice claims were given precedence in the media debate and which groups promoted them, based on the previous outline.

Table 2: Climate justice claims in SABC3, Business Day and The Mercury

Climate justice claim	SABC3 (N=18)	Business Day (N=101)	The Mercury (N=110)	Total (N=229)
Corrective justice	2	8	10	20
Distributive/ procedural justice	2	3	5	10
Intergenerational/ substantive justice	3	6	10	19
Entitlements	-	1	1	2
Ecological justice	1	1	9	11
Total claims	8	19	35	62

A total of 62 at least partly explicit justice claims were presented. It was common for more than one claim to be endorsed in the same story. Most claims were concerned with either corrective justice (historical compensation) or intergenerational/substantive justice (future opportunities). This equal approach indicates that its advocates were as concerned with the victims of the unjust distribution of historical emissions as for the well-being of future generations. The agents behind these claims can be divided evenly between politicians/negotiating leaders (primarily from Africa) and different NGOs, such as Friends of the Earth, WWF Africa, Greenpeace Africa, Oxfam, Action Aid, indigenous rights groups, Southern African Development Community (SADC) and youth delegates, including a local youth leader and Canadian youth.

With respect to distributive/procedural justice, politicians or negotiation leaders were chiefly the ones to uphold views on burden sharing, but a few NGOs also presented these views. Not surprisingly it was Saudi Arabia that presented entitlement demands. Ecological justice demands largely appeared in *The Mercury*, promoted by NGOs. Overall, this shows the importance of civil society organisations in promoting climate justice claims, but also that politicians used them either rhetorically (intergenerational) or strategically (corrective justice) to position themselves in the negotiations. In accordance with previous findings under 'Main topic', *The Mercury* presented the biggest number of climate justice claims, which were provided by a rather diverse group of agents. In *SABC3* there were seven additional storylines that drew upon a general activist frame without presenting a specific claim. For example, *SABC3* repeatedly reported on the Greenpeace activists that were deported.

(d)   Lexical choices

Table 3 summarises my findings from the lexical analysis of the full body of newspaper texts. I have chosen to contrast market-oriented language with humanitarian-oriented language in order to unpack patterns of attention, values, or world-views that were communicated. What struck me most were the similarities in the choice of language rather than the differences across three very distinct news media that targeted various audiences. 'Energy' was a buzzword across all three outlets, which popped up at least once on average in each story. The majority of the hits on energy appeared in combination with clean, green, renewable and efficiency, drawing on a technological discourse in which a switch to renewable energy would reduce emissions. Furthermore, scanning the texts for common executive and financial language, such as business, investments, management, efficiency, sustainable, green economy, and economic growth showed the great weight given to economic and administrative policies to solve the climate problem. Both newspapers produced very similar results, in which they tended to address the climate problem in terms of economic policy and technical innovation.

Nevertheless, it appears that *Business Day* leaned more often towards business as agents (refer to previous discussion) and as such indicated that business will play an important role in responding to climate change. *The Mercury* turned to concepts such as investment, management, and efficiency, which were tied to development of green energy/technology, waste/water management, and water/energy efficiency, and the like. *SABC3* seemed to be less concerned with the business role in the solution, while being more attentive to macro issues such as the transformation to a low-carbon economy and considerations surrounding employment. But the language in the newspapers also reflected a reasonable interest in a green economy and how mitigation actions might affect employment, including job opportunities.

Table 3: Lexical choice in *SABC3*, *Business Day* and *The Mercury*

Absolute occurrence (average frequency per text in brackets)	SABC3 (N=18)	Business Day (N=101)	The Mercury (N=110)	Total (N=229)
Energy	18 (1)	133 (1,3)	120 (1,1)	271
Technology	3 (0,2)	17 (0,2)	39 (0,4)	59
Business	0 -	42 (0,4)	55 (0,5)	97
Investment(s)	3 (0,2)	23 (0,2)	32 (0,3)	58

Management/ efficiency/efficient	2 (0,1)	42 (0,4)	39 (0,4)	83
Employment/ unemployment/jobs	7 (0,4)	29 (0,3)	27 (0,3)	63
Green economy/low carbon economy	10 (0,6)	15 (0,2)	15 (0,1)	40
Sustainable/ sustainability	3 (0,2)	18 (0,2)	35 (0,3)	56
Economic growth	0 -	10 (0,1)	2 (0,02)	12
Market oriented language in total	**46 (2,5)**	**329 (3,3)**	**364 (3,3)**	**739**
Poverty	4 (0,2)	9 (0,1)	10 (0,1)	23
(In)equality/ inequalities, equity, equitable	0 -	5 (0,05)	2 (0,02)	7
(Climate) Justice/ injustice	3 (0,2)	4 (0,04)	20 (0,2)	27
Fair outcome/ agreement/deal/ solution	1 (0,05)	5 (0,05)	6 (0,05)	12
Future generation/ children*/ grandchildren	2 (0,1)	7 (0,1)	9 (0,1)	18
Poor, poorer, poorest (ref nations, people)	4 (0,2)	27 (0,3)	30 (0,3)	61
Rich	3 (0,2)	29 (0,3)	32 (0,3)	64
Human rights	0 -	1 (0,01)	1 (0,01)	2
Human, humanity, humanitarian	2 (0,1)	22 (0,2)	36 (0,3)	60
Humanistic oriented language in total	**19 (1)**	**109 (1,1)**	**146 (1,3)**	**274**

*   The word 'children' is used in many contexts; only when it is used to refer to children's future is it included in this count.

The challenge of moving towards a sustainable growth path was also defined by the social context of inequality and pervasive poverty.[79] Vocabulary such as 'poverty' was frequently used in line with what could be expected. In comparison, the use of 'inequality' was essentially absent, which indicates that climate change was not debated in terms of the hot political issue of social inequalities in South Africa. The choice of vocabulary largely reflects the fact that poverty is a less sensitive issue while equality assumes a redistribution of resources. This point is underscored in this book by Kjersti Fløttum and Øyvind Gjerstad who investigated the distribution of keywords in the National Climate Change Response White Paper. They concluded that while the initial situation of poverty was recognised, it was not treated as a problem of equity.[80]

79   See Economic Development minister Ebrahim Patel quoted in *Business Day* (13 December 2011).
80   K Fløttum & Ø Gjerstad 'The role of social justice and poverty in South Africa's National Climate Change Response White Paper' (2013) in this volume.

In line with previous findings, 'climate justice' was most prevalent in *The Mercury*, while *Business Day* used the equivalent concept sparingly. As previously stated, climate justice demands are often fronted by civil society organisations, which were given more space in *The Mercury* compared to *Business Day*. The occurrences of 'climate justice' in *The Mercury* were distributed in nine articles, of which more than half reported on planned campaigns by activists, for example climate train, climate justice now, fait rally etc '"Stadium of angels" full of hope for conference'; 'We haven't borrowed the Earth from our children; we've stolen it'.[81]

Climate justice typically referred to the demands of poor and rich countries, of South and North, in terms of a fair global solution. Even though climate justice was not always expressed explicitly in a text it still might be implied, for example by calling attention to the Green Climate Fund (GCF). The GCF was one of the core issues in the negotiations and was reflected in the South African media, particularly in *Business Day*. With the exception of 'climate justice', the relative occurrences of humanistic-oriented language were rather similar across the three outlets. 'Poor' and 'rich' in this context related to the contrast between poor and rich nations. The relative high and even occurrences of this duality reflects the divide between the South and the North, particularly in the international negotiations, in the coverage of COP17.

Nevertheless, 'human rights' as a term was almost non-existent in the mediated stories that addressed climate change during COP17. Only one article in *The Mercury* explored human rights in the context of climate change in the coverage of a new position paper by the Ecosystems and Livelihoods Adaptation Network (ELAN):

> ELAN is creating a global network of scientists, policy makers and practitioners dedicated to supporting the integration of sound ecosystem management and human rights into adaptation policies, plans and programmes, especially in the world's most vulnerable countries.[82]

Business Day also had one article that linked climate change to human rights, which addressed the implications of tar sands production for indigenous people and culture in Canada:

> 'In a few years, they will no longer be able to hunt or fish. No longer will they be able to practice their traditional ways, and this affects their human rights and the rights of indigenous people,' said Mohawk from Ontario.[83]

In other words, the two articles that framed climate change as a human rights issue were based on foreign sources. However, lexical choices such as 'human, humanity or humanitarian', which might reflect the fact that climate change is a fundamental civilisation issue, were found in texts across all outlets. Even though a humanistic perspective was strongest in *The Mercury*, it was also evident in *Business Day* and *SABC3* evening news.

A distinct pattern of attention emerged during this lexical examination. First of all, regardless of the medium (newspapers v TV), the choice of vocabulary

---

81   Environmental activist Braam Malherbe quoted in *The Mercury* (28 November 2011).
82   *The Mercury* (6 December 2011) pullout section.
83   *Business Day* (7 December 2011).

seemed to reflect a strong optimism that technology would solve the climate problem. Second, climate change was to some extent portrayed as a holistic challenge in which it was linked to energy security, poverty alleviation, and the need for creating as well as protecting jobs. These findings support the idea that for emergent economies like South Africa, the transformation into a low-carbon society and a more sustainable industrial pathway is of key importance (for example, reflected in South African policy documents such as the National Climate Change Response White Paper), but not feasible without financial help from rich industrialised countries (for example, South Africa's NAMA). Finally, the lexical analysis shows that a market-oriented language outweighed the language of a humanitarian outlook by a factor of three. In particular, equity and human rights, which reflect a concern for those in current generations who are disadvantaged, were not treated as problems in the news coverage of the international climate negotiations.

VII   CONCLUDING DISCUSSION AND FURTHER QUESTIONS

The media outlets in this study made climate change a priority during COP17, and as such brought greater attention to climate change and to some extent served as an arena for promoting climate justice demands. However, there were distinct differences between the outlets, in line with what could be expected according to the media outlets' profiles and target audiences. *The Mercury* had the most expansive coverage of a wide range of *topics* and issues, tended to be more sensitive to social and environmental claims from civil society, and wrote with a sharper pen. *Business Day* focused on a more limited set of topics, actors, and discourses that fit the newspaper's profile, and presented many issues in economic and management terms in their response to climate change. *The Mercury's* diversity of topics may reflect a broader readership than *Business Day.*

While *SABC3* probably had the most diverse audience among the outlets, its television format limited the overall number of news items and provided less opportunity to go into depth on issues. This in turn demanded tougher editorial decisions as to what to broadcast. *SABC3*'s political ties also appeared to affect the prioritisation of topics and actors, which were more political than the other media outlets. The analysis also illustrated that *SABC3* drew heavily on an action-oriented frame, which gave visibility to activists' demands or protests, but with little substance. This can be partly explained by the incorporation of a commercial orientation and the fact that television is a visual medium.

The analysis also showed that it was relatively common for the South African media to link climate change to other global/development challenges, such as poverty, food security, vulnerability, and adaptation. About a fifth of the stories drew on a development discourse. As such, it can be assumed that hosting COP17 helped to reframe climate change as a holistic sustainable development challenge in the South African context, including the need to balance mitigation with energy security, employment creation, and poverty alleviation.

Top politicians and chief negotiators, who are key actors at the international climate negotiations, were given the greatest power to define what climate change was all about in the public debate. However, NGOs/activists constituted the second largest group of *agents* in *The Mercury* and *SABC3* and the third largest group in *Business Day*, which indicates that they were quite successful in conveying their messages. While civil society activists and organisations often played a prominent role in *The Mercury* stories, the same was true for business representatives in *Business Day*, which led to very different kinds of stories and frames for understanding climate change. While the former often portrayed climate change within a morality frame, where demands for justice were promoted, the latter was more concerned with how business could play a vital part in the solution. However, it is essential to underline that connections between media messages, policy decision-making, and behavioural change are far from straightforward. Media coverage does not determine engagement; rather it shapes possibilities for engagement.[84]

A climate justice movement has massive support from advocacy NGOs and religious groups all over the world, which at least in part explains the resonance of climate justice demands in the media. However, efforts to put a human face on climate change and to link to local communities were rarely reflected in the media output. Overall, *victims* of climate change were referred to in non-personalised broader terms, such as Africans or poor countries or poor people, which limited the possibility to identify and engage with them. The complex nature of climate change appeared to make it difficult for journalists to dress up their stories in a human-interest narrative, which probably also weakened the links between protection of human rights and climate change.

Framing climate change as a *human rights issue* was virtually absent in the media material during COP17. One reason might be the nature of this relationship as complex and indirect, while mass media prefers simple and straightforward storylines. Another reason might have been the lack of sufficiently nuanced information to cover the situation of individuals and communities who experience climate impacts as infringements on their rights. This could have made it difficult for advocacy groups to promote these aspects and unlikely for the media to frame climate change as a human rights issue. Nonetheless, the value of using human rights to frame the issues, or as a discursive vehicle in storytelling, could help the media bring a human face to climate change. Real people in specific places, and their basic rights to health, food, water and shelter, are much easier for any media audience to identify with than abstract facts and figures. In other words, a human rights lens could potentially spotlight possible harms that might otherwise be missed or ignored.

Furthermore, the analysis revealed that the media outlets had a strong *technological optimism* regarding how the climate problem could be solved. This supports the idea that for emergent economies like South Africa, the transformation into a low-carbon society and a more sustainable industrial

---

84   Boykoff (note 4 above) 2; Carvalho & Burgess (note 8 above).

pathway is of key importance, but not feasible without financial help from rich industrialised countries. The lexical analysis documented the predominance of market-oriented language, which outweighed the language of a humanitarian outlook by a factor of three. The frequent use of 'energy' strengthened the impression that technology could fix the climate problem. The salience of green technologies and innovations might be explained by the fact that they are characterised by practical proposals, concrete products that are immediately useful, and they are a non-threatening option (except perhaps for fossil fuel industries).

In contrast, climate justice involves visions, demands, and humanistic values, which might be easier to use as basis for a critique rather than being operationalized into practical applications. Most justice claims were concerned with either corrective justice or intergenerational/substantive justice, which suggests there was equal attention paid to the victims of the unjust distribution of historical emissions and the well-being of future generations. However, the lexical analysis revealed that the news coverage failed to show that a sustainable growth path is also defined by the social context of inequality. 'Inequality' and 'human rights', which reflect a concern for the disadvantaged of the current generation, were not treated as problems in the news coverage. As such, the news coverage didn't encourage its readers to view climate change through a lens that showed disadvantaged urban or rural poor who risk their human rights to be further endangered by climate change impacts. The lack of human rights language might be part of a larger trend, as documented by Des Gasper, Ana Victoria Portocarrero and Asunción Lera St. Clair in this book.[85] They claim that important concepts such as human rights, future generations, humanity, and poverty have disappeared from the *Human Development Report* 2011, and have instead been replaced by the language of sustainability, green economy, in terms of solutions. In other words, they found a convergence towards a World Bank perspective. However, the picture for the media is far from black and white in this respect.

All in all, the selected media outlets in this study provided extensive news coverage of COP17, but failed to address socio-economic inequalities when responding to climate change and created limited opportunities for engagement due to the lack of a personalised narrative. Looking ahead, analyses of non-mainstream media platforms as well as a broader range of cases might improve our insights into how climate justice and right views on poverty might be addressed. Further studies could also examine how the media tackle the complexities of relevant causes and effects that are spatially dispersed across the globe as well as temporally dispersed. Finally, it has been my intention to analyse the 'political' representations of COP17 against the backdrop of shifting global geo-political power relations, which saw the rise of regions such as Brazil, South Africa, India, and China. These shifts have led to a struggle over worldviews, which might have implications for climate justice demands.

---

85   D Gasper, AV Portocarrero & AL St Clair 'An analysis of the human development report 2011:
     Sustainability and Equity: A better future for all' (2013) in this volume.

CONTESTING CLIMATE INJUSTICE DURING COP17

BRANDON BARCLAY DERMAN[*]

I INTRODUCTION[1]

Africans are the continental population most affected by, and least responsible for, the impacts of climate change.[2] As the United Nations Framework Convention on Climate Change (UNFCCC) prepared to descend on Durban for its 17th yearly Conference of the Parties (COP17) in late November 2011 many non-governmental organisations (NGOs) and social movements formulated positions around themes of justice in connection to Africa's situation and the political upheavals of 2011. Two developments within the convention importantly shaped those positions: the implementation of the Green Climate Fund (GCF), agreed during the previous COP as a means to 'scale up the provision of long-term financing for developing countries',[3] and the question of binding commitments to reduce emissions after the (planned, but now averted) expiration of the Kyoto Protocol in 2012.

In this chapter I describe mobilisation for climate justice in Durban during COP17 by groups operating at transnational, pan-African, national and local scales. I am particularly concerned both with themes that cut across the work of these groups, and with the ways in which South and southern African groups articulated relations between the global politics of climate governance and regional, national, and local conditions. In Durban, demands by civil society for ambitious and binding mitigation in developed countries, and climate finance for Africa were frequently linked with struggles for social justice, accountability, political representation, and survival, highlighting structural inequality and social division at national as well as global scales. Groups such as the Pan-African Climate Justice Alliance (PACJA) and the Rural Women's Assembly (RWA), for instance, couched demands for mitigation and finance in terms of tenuous

---

[*] This material is based upon work supported by the United States National Science Foundation under Grant No. 1129127. The author bears sole responsibility for its content.

[1] This chapter draws on direct observations to help describe the framings associated with concerns of justice during COP17. As part of a longer research project on justice, civil society and climate governance, I spent three weeks in Durban prior to and during the COP, observing intergovernmental and civil society meetings and public demonstrations. Where permitted and feasible, I recorded speeches, otherwise writing field notes. Recordings and field notes were transcribed, coded and analysed for themes, following standards of qualitative analysis, for example, L Richards *Handling Qualitative Data: A Practical Guide* (2009). Quotations and summaries of views communicated by meeting participants included here come from those transcripts. Descriptions of the Global Day of Action march and other events are additionally based on photographic and video documentation.

[2] AK Nord & K Omari (eds) 'Mobilising Climate Finance for Africa' (2011) 4 *Perspectives* 3.

[3] See <http://cancun.unfccc.int/financial-technology-and-capacity-building-support/new-long-term-funding-arrangements/>.

pastoral livelihoods and the particular capacities and vulnerabilities of African women. Both those groups also sought to increase accountability in climate governance, bringing stakeholders facing the cutting edge of climate impacts across thousands of kilometres to Durban to represent themselves and their interests. Dramatising similar commitments inside the International Convention Centre (ICC), transnational activists 'occupied' the COP in the name of direct democracy. In another example of activism grounding broadly shared principles with national context, the South African Right2Know campaign linked its domestic transparency initiative with opposition to closed-door deal-making in COP17,[4] amidst mounting critique in transnational networks of corporate influence and limits on civil society in UN bodies.[5]

Human rights conventions articulate principles that have also been marshalled in support of mitigation and adaptation finance in the name of climate justice.[6] The Human Rights Council (HRC) has acknowledged that climate change poses 'implications' for the enjoyment of rights,[7] and yet these efforts have gained little traction in the UNFCCC.[8] As the work summarised below suggests, where rights discourse appeared in mobilisation for climate justice during COP17 it often played a supporting role within messages, actions, and campaigns highlighting inequality and social divisions at international and national scales. These pointed toward the necessity of responsible national as well as international actions, and the imperative of timely (ie immediate) mitigation and finance for Africans and others least responsible for the climate impacts they face head on.

II ADVOCATING CLIMATE JUSTICE DURING COP17

(a) Transnational campaigns for climate justice in Durban

During COP17, transnational NGOs and social movement groups pursued aims of climate justice focused around a set of broadly related themes. In addition to publishing their analyses in position papers and by faster-moving means on the Internet, organisation members participated in official COP side events,[9] organised sessions at the People's Space convened at the

---

4   Right2Know called on COP delegates to 'Stop the Climate Secrets' (1 December 2011); Right2Know Campaign 'R2K Joins the Call for Access to Information for Climate Justice!' <http://www.r2k.org.za/2011/12/01/stop-climate-secrets/>.

5   See <http://www.tni.org/events/dismantling-power-tncs-reclaiming-development-alternatives>.

6   For example, Inuit Circumpolar Council *Petition to the Inter American Commission on Human Rights Seeking Relief from Violations Resulting from Global Warming Caused by Acts and Omissions of the United States* (2005); Maldives Human Rights Council *Resolution 7/23 'Human Rights and Climate Change': Submission of the Maldives to the Office of the UN High Commissioner for Human Rights* (2008).

7   JH Knox 'Linking Human Rights and Climate Change at the United Nations' (2009) 33 *Harvard Environmental LR* 477.

8   Center for International Environmental Law *Analysis of Human Rights Language in the Cancun Agreements (UNFCCC 16th Session of the COP)* 2011; A Johl & S Duyck 'Promoting Human Rights in the Future Climate Regime' (2012) 15 *Ethics, Policy & Environment* 298.

9   See <http://unfccc.int/files/meetings/durban_nov_2011/application/pdf/see_brochure_cop_17_cmp_7.pdf>.

University of KwaZulu-Natal (UKZN),[10] and marched on the Global Day of Action.[11]

A number of campaigns linked the livelihoods and practices of small-scale farmers in Africa and elsewhere with the threats of climate change. La Via Campesina, for instance, argued 'industrial agriculture heats up the planet, farmers are cooling it down', while Oxfam and its partners demanded, 'grow food, not emissions!'[12]

Other groups denounced false solutions, corporate influence, and dirty energy projects. Many issued calls to 'leave the oil in the soil',[13] and critiqued soil carbon markets.[14] Organisations presented watchdog reports on bank investments in fossil-fuel energy projects and the risks of crime in Reducing Emissions through Deforestation and Forest Degradation (REDD) programmes at COP side events,[15] and held panels devoted to 'dismantling the power of TNCs [transnational corporations] and reclaiming development alternatives' at UKZN.[16]

The Peoples' Space hosted a conference on eco-socialist theory and practice[17] organised by South African group, the Democratic Left Front (DLF).[18] Presenters there articulated solutions to the climate crisis based in strategies of commoning and the recognition of intrinsic value in ecological integrity, the analyses of social movements and communities in the global south, linking capitalism's exploitation of fossil-fuels with its discounting of women's labour, and provisions for just transition based in democratic control of oil resources.[19]

A transnational coalition for the Rights of Nature[20] participated in the eco-socialism conference and several other events in Durban.[21] At UKZN, former chief negotiator for the Plurinational State of Bolivia, Pablo Salon, described the relevance of the Proposed Declaration of the Rights of Mother Earth[22] for international climate policy. Salon argued that Rights of Nature provide

---

10   See <http://cop17insouthafrica.wordpress.com/2011/11/08/ukzn-to-accommodate-alternative-co p17-civil-society-events/>.

11   See <http://cop17insouthafrica.wordpress.com/2011/12/02/global-day-of-action-rallying-call-du rban-climate-march/>.

12   Author's photographs of march placards, and online coverage by La Via Campesina <http:// viacampesina.org/en/index.php/actions-and-events-mainmenu-26/-climate-change-and-agrofuels-mainmenu-75/1126-via-campesina-at-cop17-in-durban-industrial-agriculture-heats-up-the-planet-farmers-are-cooling-it-down>; and Oxfam Canada <http://twitter.yfrog.com/ ntncgfgj>.

13   For example, CDM Watch in its series 'Watch this! Progress and Gossip about Carbon Markets at COP 17' <http://carbonmarketwatch.org/wp-content/uploads/2011/12/watch-this-311.pdf>.

14   ActionAid 'Fiddling with Soil Carbon Markets While Africa Burns...' (2011) <http://www. ms.dk/en/uganda/shared/fiddling-soil-carbon-markets-while-africa-burns>.

15   UNFCCC (note 9 above).

16   TNI (note 5 above).

17   See announcement <https://durbanclimatejustice.wordpress.com/2011/11/26/ecosocialism-con ference-1-december/>.

18   See <http://www.democraticleft.za.net/>.

19   These brief summaries are based on my own notes. The most comprehensive summaries I have found online are by Saul <http://ecosocialisthorizons.com/2011/12/the-movement-has-begun/> and Dawson <http://www.socialtextjournal.org/periscope/2011/12/amandla-for-eco-socialism.php>.

20   See <http://therightsofnature.org/>.

21   See <http://therightsofnature.org/events/cop-17-rights-of-nature-events-in-durban-sa/>.

22   World Peoples Conference on Climate Change and the Rights of Mother Earth *Proposal: Universal Declaration of the Rights of Mother Earth* (2010).

a basis for transition from present day 'apartheid against nature', toward a socially and environmentally just society founded on a non-exploitative model of development. Moreover, in the international context the Rights of Nature offers an alternative to the commodification of nature through 'green economy' initiatives ascendant in UN institutions and transnational business circles. Audience members at UKZN engaged with Salon and each other in discussion of Rights of Nature principles at other political scales, and in the context of deep social division, poverty, and the unrecognised or realised rights of people.[23] 'I'm worried about the discourse of the rights of nature,' said one observer, wondering whether its terms could be marshalled to support poor people in South African cities struggling for affordable access to water. Acknowledging those concerns Salon underlined that Rights of Nature constitute 'part of the solution' and that redistribution at national levels remains essential. 'Rights of Nature is an appeal to think in a much broader perspective,' Salon argued, however. In recognition of global limits to development based in fossil-fuel extraction 'we have to act as a world, not individual countries'.[24]

(b) Occupying COP17

Assemblies, speeches, vigils, and informal gatherings occurred on an ongoing basis at the Speakers' Corner opposite the ICC under the banner of Occupy COP17:

> a forum for those who wish to discuss and implement real and equitable solutions to climate change, with climate justice at the heart. It is open to all, operating on the principles of inclusiveness, openness, non-hierarchical organising and consensus decision-making.[25]

To that end, Occupy COP17 used horizontal modes of communication common to Occupy protests on Wall Street and elsewhere.[26]

The development of shared positions through the open forums at Occupy COP17 was seeded with the People's Agreement from Cochabamba[27] and framed around imperatives of direct democracy and opposition to finance capital and corporate power:

> the very same people responsible for the global financial crisis are poised to seize control of our atmosphere, land, forests, mountains and waterways. They want to institute carbon markets that will make billions of dollars for the elite few, whilst stealing land and resources from the many.[28]

Assemblies at Speakers' Corner developed these starting points, advocating 'local level solutions including the idea of integrating mother earth rights in local bylaws', 'equity between humans and nature', integrating climate-based

---

23   Direct observation, as described in note 1 above. As the exchange described suggests, audience members raised questions around the apparent priority of 'nature's' rights over those of people to access and use natural resources such as water.
24   Author's summary. See announcement for and video documentation of Salon's talk at UKZN <http://ccs.ukzn.ac.za/default.asp?11,22,5,2668#Rights%20of%20Nature>.
25   See <http://occupycop17.org/about/>.
26   Direct observation. Participants in Occupy gatherings typically use, for example, the 'human mic' and general assembly format. See <http://www.youtube.com/watch?v=xIK7uxBSAS0>.
27   See <http://occupycop17.org/possibilities/>.
28   See <http://occupycop17.org/about/>.

activism 'with the need for economic change, gender, and a variety of other factors', and asserting 'people power over the corporate influence that is trying to promote false solutions and take control of those false solutions with intellectual property rights'.[29]

Occupy COP17 issued a letter of support for the International Day of Action for Human Rights, drawing a parallel between the immediate reality of climate impacts in Africa and the longer time scale of inter-generational injustice associated with continuing anthropogenic warming:

> Our planet is changing, and with it the story of human rights. Here in Africa, the river beds are already drying and the seedlings that we watered for the future, are wilting. We do not rob water from the cups of others, we divert the streams. Climate change is the tyranny of the present over the rights of the future. In Africa, the future is already here.[30]

At the closing of the COP, Occupy COP17 voiced support for the analysis of Climate Justice Now!, which characterised the summit's outcome as a 'crime against humanity', wherein 'the richest nations have cynically created a new regime of climate apartheid'.[31]

During COP17, occupiers at the Speakers' Corner provided a visible space of dialogue and protest and a nexus for groups critical of corporate interest in the negotiations. There was an international cast to the set of organisers, participants and guests at Occupy COP17, which included members of 350.org, and UNFCCC delegates from the Small Island Developing States group, but members of South and southern African civil society groups such as the RWA and DLF also participated in assemblies there, and used the space for gatherings.[32]

Prominent actions involving accredited observers inside COP17 also took shape through the language and practices of Occupy protests. On 9 December observers and a handful of party delegates blocked a hallway and moved gradually toward a plenary until they were stopped by COP17 security officials. Many surrendered their accreditation as they were escorted from the convention centre. Like those who marched on the Global Day of Action, these protestors within the ICC chanted 'listen to the people, not the polluters'.[33] On the same day Anjali Appadurai ended her statement on behalf of global youth to the COP17 plenary with the 'mic check' that structures dialogue at Occupy events, and then led observers congregated at the back of the room in a collective call for climate action and 'equity now'.[34]

---

29   See <http://occupycop17.org/ga-notes/>.
30   See <http://occupycop17.org/2011/12/10/occupycop17-supports-international-day-of-action-for-human-rights/>.
31   See <http://occupycop17.org/2011/12/11/occupy-cop17-supports-cjn-statement-at-the-conclusion-of-cop17/> quoting <http://www.climate-justice-now.org/2011-cop17-succumbs-to-climate-apartheid-antidote-is-cochabamba-peoples%E2%80%99-agreement/>.
32   Direct observation. The Speakers' Corner was the centre of an impromptu march involving those groups on the day before the global day of action. The website for the gathering also provided video statements by international activists <http://occupycop17.org/videos/>.
33   Direct observation. See Takver 'OccupyCOP: Hundreds Protest inside UN Climate Venue in Durban as Talks Draw to a Close' *Climate IMC* (2011) <http://www.indybay.org/newsitems/2011/12/10/18702334.php>.
34   See A Goodman 'On Climate Change, the Message is Simple: Get it Done' *The Guardian* (14 December 2012) <http://www.guardian.co.uk/commentisfree/cifamerica/2011/dec/14/durban-climate-change-conference-2011>.

(c)   The Pan-African Climate Justice Alliance

Prior to and throughout Durban's COP17, PACJA[35] analysts articulated a vision broadly shared by transnational climate justice networks, demanding fair climate finance combined with deep and binding emissions cuts in developed countries. Speaking on the former prior to COP17, Michelle Maynard of PACJA and the Centre for Civil Society (CCS) at UKZN, argued:

> The principles are simple: providing climate finance is a legal and moral obligation for rich countries ... It is a legal obligation included in the UN Convention that every country, including the United States, agreed to. It's a moral obligation arising from rich countries' climate debts – debts they owe from overusing their fair share of the atmosphere and from causing climate change and climate change harms ...
>
> Funding should be from public sources, new and additional to Official Development Assistance (ODA). Rich countries should not shirk from their responsibilities by anthropocentrically commodifying nature through the market.[36]

PACJA also led what was probably the largest pan-African mobilisation effort to target COP17. The 'Trans African Climate Caravan of Hope', brought approximately 300 activists, members of civil society, scientists, farmers and journalists to Durban.[37] The caravan crossed some 7,000 kilometres and ten countries raising awareness and collecting travellers and messages.[38]

A PACJA spokesperson described the goal behind the caravan at a meeting on the UKZN campus:

> the climate justice movement cannot survive if we don't involve farmers, if we don't involve women, if we don't involve those who are impacted heavily by climate change. Those are the pastoralists, and people who are at the forefront ... of this problem and that was the spirit of this caravan.

By bringing individuals and messages from the forefront to Durban, PACJA sought, against the odds, to push for an agreement that would be 'responsive to African realities'.[39] Many along the caravan route added their signatures to a petition demanding that the COP:

1. Keep Africa and the world safe and prevent catastrophic climate change. Exert pressure on developed countries and ensure that they sign up to legally binding commitments that reduce emissions and limit global warming to well below 1.5°C.
2. Share the effort of curbing climate change fairly. Demand domestic emission reductions by developed countries that are commensurate with science and equity, and enable a just transition in all countries.
3. Ensure polluters not the poor must pay. Developed countries must honour their obligations and pay at least 1.5% of their GNP to help the poor adapt and develop cleanly and sustainably.[40]

---

35   See <http://www.pacja.org/>.
36   Maynard quoted in A Rafalowic 'Markets Skew Climate Talks to Favour Rich | Civil Society Condemn Offsets and World Bank role in Climate Finance' *Climate Justice Aotearoa* (n.d.) <http://www.climatejusticeaotearoa.org/2011/04/05/markets-skew-climate-talks-to-favour-rich-civil-society-condemn-offsets-and-world-bank-role-in-climate-finance/>.
37   H Gersmann & J Vidal 'Q&A: Durban COP17 Climate Talks' *The Guardian* (28 November 2011) <http://www.guardian.co.uk/environment/2011/nov/28/durban-cop17-climate-talks>; 'Climate Campaign Travelling Across Africa to COP17' <http://www.mrfcj.org/news/2011/climate_campaign_travelling_across_africa_to_cop17.html>.
38   ActionAid '7000km in 17 days to COP17' <http://www.actionaid.se/en/activista/shared/7000km-17-days-cop17>.
39   From author's direct observation, transcripts and recordings.
40   The petition is summarised in ActionAid (note 38 above).

(d)   The Rural Women's Assembly

The RWA of the Southern African Development Community (SADC) region, a coalition of international women's NGOs, convened for the third time in Durban during COP17, bringing together over 650 women from nine countries.[41] As Samantha Hargreaves describes, the regional and national chapters of the RWA seized COP17 as an opportunity to build group cohesion and recognition, and to increase its constituents' knowledge of climate issues. They pursued a groundbreaking process of organisation and advocacy combining grassroots and policy engagement, particularly around the drafting of the South African Climate White Paper:

> ... over the period of May to mid-December 2011 [the RWA] undertook an impressive number of workshops, awareness-raising activities, marches, demonstrations and two assemblies, and wrote no less than five different memorandums, policy submissions and a dozen letters to different government departments.[42]

In Durban during COP17 their highly visible presence had an energising effect on other groups.[43] In the main memorandum they issued during COP17, the RWA called for specific, ambitious, binding commitments by developed countries and 'adequate public finance to meet Africa's mitigation and adaptation needs'.[44] They also called attention to the position of rural southern African women facing climate change under what Hargreaves calls the 'triple burden of race, class and geographical marginalization':[45]

> Rural women across Southern Africa are already reporting 20 per cent decreases in food production, and current trends tell us that if we fail to take action now, by 2020 we will have seen a 50 per cent loss in crop yields in our region. We produce 80 per cent of the food consumed by households in Africa. In the absence of support for us, we believe that local and national food security will be deeply threatened.
>
>   We ask that you properly recognise women's critical role in fighting climate change and protecting livelihoods and the environment. Equal rights to land and natural resources is critical to fight climate change. The Rural Women's Assembly asks that governments implement the principle of 50/50 land to women through national programmes of land redistribution and agrarian reform.[46]

By Hargreaves' analysis, the RWA's work prior to and during COP17 appreciably furthered the group's growth in terms of membership, experience and internal cohesion.[47] Their efforts to realise their rights and interests had little effect, however, on the COP decisions or the South African White Paper:

41   S Hargreaves 'COP 17 and Civil Society: The Centre did not Hold' Institute for Global Dialogue *Occasional Paper* 64 (2012).

42   Ibid 13.

43   Direct observation, for example expressed through the consideration and respect of members of other groups in a multi-group preparation session the evening before the Global Day of Action.

44   RWA 'Memorandum from the Rural Women's Assembly to President Zuma and Minister Nkoane-Mashabane' <http://ruralwomensassembly.wordpress.com/cop17/memorandum/>.

45   Hargreaves (note 41 above) 9.

46   RWA (note 44 above).

47   The group later took its socio-ecological activism to Rio de Janeiro for the Conference on Sustainable Development <http://www.aljazeera.com/indepth/features/2012/06/2012622175745190650.html>.

The failure to set an adequate emissions reduction target (required to keep temperature increases below the scientifically accepted 1.5 degree standard) will have dramatic consequences for rural women, the major subsistence and small-scale producers, and their offspring in just a decade or two.[48]

The interests of the RWA, and the many constituents its member organisations represent, thereby remained marginal in national and international political processes.

III   WORKING FOR CLIMATE JUSTICE IN DURBAN, RIGHTS OR NO

Durban has been the locus of formative work for climate justice,[49] and local and national organisations mounted a range of efforts during COP17.

At UKZN and in the streets South African trade unions, social movements and environmental groups promoted the launch of a campaign for One Million Climate Jobs based on research into South African economic conditions and climate vulnerabilities. The campaign calls for 'a just transition to a low carbon economy to combat unemployment and climate change', through the creation of jobs that:

(1) reduce the amount of greenhouse gasses we emit, to make sure that we prevent catastrophic climate change; (2) build our capacity to adapt to the impacts of climate change (for example, jobs that improve our food security); (3) provide and secure vital services, especially water, energy and sanitation (this includes reducing wasteful over-consumption).[50]

The cross-cutting character of that campaign typified the work of a set of national and local groups including the DLF, groundWork,[51] and the South Durban Community Environmental Alliance (SDCEA),[52] who combined social and environmental justice positions at events in the Peoples' Space, in COP17,[53] through mobilisation, and through 'toxic tours' of local communities impacted by refineries.[54] Like PACJA and the RWA, DLF and groundWork organisers were heavily involved in linking stakeholders 'outside' the COP17 process with the debates and conditions of uneven power on the 'inside', and linking community-based concerns with critical analysis of global governance.[55]

In light of UNFCCC reticence on the relationship between human rights and climate change,[56] it is understandable that human rights discourse played a supporting, largely political role in mobilisation for climate justice during COP17. Advocates focused on those structures of opportunity within the

---

48   Hargreaves (note 41 above) 15.
49   See interventions by the Durban Group for Climate Justice <http://www.durbanclimatejustice. org/> and as described in P Bond *The Politics of Climate Justice: Paralysis Above, Movement Below* (2011).
50   See <http://climatejobs.org.za/>.
51   See <http://www.groundwork.org.za/index.html>.
52   See <http://www.sdcea.co.za/>.
53   As in SDCEA's Desmond D'Sa statement at a COP side event <http://www.youtube.com/ watch?v=3dBm6qE24HM>.
54   Captured in a video document <http://www.youtube.com/playlist?list=PLVS3GQE0TK00tXCF mZRzQX0Pa2iTofq4M>.
55   Summary of direct observation. DLF organisers, for example, worked to mobilise ex-urban constituencies for the Global Day of Action, while groundWork representatives coordinated panel discussions at UKZN on the relationship between social movements and climate negotiations.
56   Johl & Duyck (note 8 above.)

convention that might have enabled addressing urgent needs (implementing the GCF to deliver climate finance; and ambitious, binding mitigation targets). They drew on existing language about responsibility and legal obligation, and a revivified global movement for democracy, connecting them with accelerated climate impacts, poverty, and inequality in Africa. Local, South, and southern African groups linked these themes with the conditions of the rural and urban poor, who face livelihood threats exacerbated by ongoing climate changes that remain largely unaddressed by dominant responses, and may lack access to meaningful representation or legal recourse.

Much of the work analysing climate change impacts in relation to human rights recognised under international law[57] reflects Mary Robinson's view that '[h]uman rights law is relevant because climate change causes human rights violations', and that 'human rights make clear that government obligations do not stop at their own borders'.[58] The impact of efforts based on these analyses, however, has been limited by competing understandings of human rights as the responsibility of states to citizens only within their borders,[59] and by legal theories of responsibility which make it difficult to attribute harms associated with climate impacts to specific emitters, since emissions are dispersed throughout the atmosphere, and impacts are experienced through 'environmental' change.[60] Put more abstractly, international divisions of global space, and legal division between human action and ecological impacts have thwarted formal climate change rights claims. By linking the topics of debate in the UNFCCC with local, national, regional, and global inequalities in harm, access to resources, and political power, climate justice mobilisation during COP17 highlighted social division as a third systemic condition underlying climate injustice, against which legal rights may also be, on their own and as yet, inadequate tools. The 'wicked-ness' of climate change as a social problem[61] is continually illustrated by the failure of political processes to address the multiple issues of justice it entails.[62] These demand responsible and accountable action in the immediate term, both across national borders and within them.

---

57   For example notes 6 above and Knox (note 7 above); International Council on Human Rights Policy (ICHRP) *Human Rights and Climate Change* (2009); S McInerney-Lankford 'Climate Change and Human Rights: an Introduction to Legal Issues' (2009) 33 *Harvard Environmental LR* 431; M Limon 'Human Rights and Climate Change: Constructing a Case for Political Action' (2009) 33 *Harvard Environmental LR* 439.

58   ICHRP ibid iii–iv.

59   Knox (note 7 above).

60   See, for example, M Chapman 'Climate Change and the Regional Human Rights Systems' (2010) 10(2) *Sustainable Development Law & Policy* 10; and Knox (note 7 above).

61   R Lazarus 'Super Wicked Problems and Climate Change: Restraining the Present to Liberate the Future' (2009) 94 *Cornell LR* 1153; K Levin, B Cashore, S Bernstein & G Auld 'Playing it Forward: Path Dependency, Progressive Incrementalism, and the "Super Wicked" Problem of Global Climate Change' (2010) *International Studies Association 48th Annual Convention* vol 28 <http://environment.research.yale.edu/documents/downloads/0-9/2010_super_wicked_levin_cashore_bernstein_auld.pdf>.

62   SM Gardiner 'Climate Justice' in JS Dryzek, RB Norgaard & D Schlosberg (eds) *The Oxford Handbook of Climate Change and Society* (2011).

QWASHA! CLIMATE JUSTICE COMMUNITY DIALOGUES COMPILATION VOL 1: VOICES FROM THE STREETS

MOLEFI MAFEREKA NDLOVU

I INTRODUCTION

The Constitution of the Republic of South Africa, 1996 expresses a commitment to environmental rights and guarantees the rights of all to participation and expression.[1] Yet the vast majority of ordinary South Africans have not mobilised around these rights in the context of climate change. Why is this the case?

Many writers have suggested that the low level of participation by activists and members of poor communities is due to the lack of knowledge about and interest in the climate change phenomenon. However, this research suggests that this may not necessarily be the case, since marginal communities have already started to feel the impact of climate change in their daily lives, like the high and rising costs of basic food staples such as flour, maize, vegetable oil, sugar and other staple grains, as well as the rising cost of energy reflected by increases in energy prices (25 per cent annually since 2008 in South Africa); this, coupled with the increase in droughts and floods that have led to the general reduction in food supply and rising uncertainty about the future.

This chapter – and the Qwasha![2] Project on which it is based – used the occasion of the United Nations Framework Convention on Climate Change (UNFCC) Conference of Parties 17 (COP17), which took place in Durban during November and December 2011, as an opportunity to try to investigate the relative absence of rights-based climate change mobilisation. It did so by dialoguing with various members of South African civil society about their understanding of the climate change phenomenon, their views on the formal COP17 process, as well on the role of the official civil society space hosted by the Committee of 17 (C17) civil society organisations in facilitating participation and critical expression for ordinary people outside of the officialdom that has come to characterise international events such as COP17.

The Qwasha! Project uses a web platform[3] as a depository for materials gathered after setting up, training and facilitating local collectives to run local archival hub stations across eThekwini municipal area. Initial networks have

---

1   Constitution Chapter 2 s 24 <http://www.info.gov.za/documents/constitution/1996/96cons2. htm#24>; also see <http://www.climateresponse.co.za/home/gp/1>.

2   Qwasha is an isiZulu word [v/i], imperative, singular of ukuQwasha (v/i) to lie awake without sleep, very alert. Used in everyday language to denote becoming conscious of one's surroundings.

3   See <http://www.qwasha.org.za>.

been created and interest is high in the 15 Durban communities: four in the South Durban basin (Umlazi, Chatsworth, Wentworth, Folweni); four in the centre (Durban CBD, Umbilo, UKZN, Overport); and seven in the northern parts of the metropolitan area (Inanda, uMzinyathi, KwaMashu, Marrianrige, Clermont, KwaNgcolosi, Durban North).

Qwasha! *Climate justice community dialogues compilation* vol 1, is essentially a series of 'pavement-broadcasts', made on the streets, based on audio conversations, recordings of songs, life stories and interviews by the Qwasha! collective-in-the-making. The bulk of observations, interviews and conversations took place between November and December 2011. Typically conversations were initiated by an open-ended question such as: 'Can you tell me what you think COP17 is about?' Based on the response, follow-up questions were then asked. This approach proved useful to the extent that no two interviews were responded to in the same manner; hence a great variety of responses and perceptions were generated.

The audio recordings consist of a total of 55 semi-formal interviews and 17 audio recordings of songs and conversations. Among those interviewed were: three local development practitioners/members of local development organisations; four members of C17 organisations; 10 women activists (especially rural (6) and elderly women (3), but also three young women respondents) who gave perspectives on how climate change/Cop17 impacted women's lives; four respondents from the 'COP17 volunteers' or the 'Green Bombers', as they came to be called; and 25 random interviews, which included stories of fisher folk and members of subsistence farming communities, casual workers within the services, securities and cleaning sectors, taxi drivers, and informal traders at the Warwick Junction Market Precinct and migrant communities. Three of the interviews were with official participants within COP17, nine interviews and conversations were with activists attending the alternative civil society spaces organised by C17, and eight interviews and conversations were with artists and counter-culture activists to gather views on how culture and critical awareness can inform and critique the politics of climate justice. The above, taken together, represent a sample of the demographic variables including, for example, gender, age, literacy level, class position and race.

So what are the participants' views on climate change? How has the COP17 process influenced their understanding and capacity to make a change from an individual, group and community level, if at all? What are the messages that are filtered down to the streets where ordinary people dwell? Can ordinary community voices find relevance in these global negotiations – or in the organised civil society activities?

II   QWASHA! VOL 1: VOICES FROM THE STREETS

The views captured in this series can be regarded as a reflection of micro perspectives of how ordinary people understood, framed and articulated a narrative about the global climate change phenomenon. This narrative helps provide a counter-balance between official and formal civil society discourses, which are framed within a 'summit' discursive paradigm that were in circulation during COP17 in Durban. Some scholar activists have held that the COP17 'summit' can be read as a platform for nation states to negotiate global issues. As a fellow researcher, Felix Platz, put it at a Centre for Civil Society (CCS) seminar presentation: 'an arena in which different actors construct, articulate their issues and discourses'.

Below we have selected and rendered to written text, a few quotes from the audio interviews, songs and retellings, which seem succinctly to capture the essence of the entire audio collection, whilst taking cognisance of the need to avoid bias and to be demographically representative of the total sample of views shared.

When asked about her perceptions of what COP17 is about, Emily Dlomo, a long-time gender and community organiser from KwaMashu, whom we met on the streets of Durban during the march on the Global Day of Action (3 December 2011), said:

> People are asking what is COP17, not much has been done on the ground to show people the uselessness of these negotiations, out of ten people only three could tell you what is climate justice. Only a select few in city based NGO [non-governmental organisation] have an idea. Even as we march here today, I can promise you that all these people standing on the sidewalks have no idea whether we are marching in favour of climate change or against it. Much more needs to be done at community level where talk about the effects of climate change can be explained based on people's lived experiences. 87% of arable South African land that was stolen by the apartheid regime away from its original users is yet to be returned, the new democratic regime has only been able to redistribute ... like 4%, that means we do not have enough land and the little land we have has been poisoned by the same industrialists and large-scale commercial farmers. Unless these facts are at the centre of climate justice agenda, it will remain an area colonised by middle-class environmentalist.

Emily Dlomo is advanced in age, but her spirit is energetic, having been active since the 1970s in the struggle against apartheid and the oppression of women; she is still fighting to this day. She shares her uncertainties about the future of South Africa, especially in this day and age where the challenges seem so much more, yet the enemy is not as clearly defined. Emily Dlomo feels that rural people, especially women, have been neglected by the current political dispensation; yet the same rural communities hold the answers to the challenge of climate change.

Regarding rights, Nobuhle Sokhela is frustrated. Nobuhle, who matriculated in 2000, has two children and lives with her mother and three unemployed brothers in Embumbulu in the south of Durban, stated:

> It seems black people and poor people in this country do not count for much, we have been denied our rights to a better education, we have been denied the opportunity to get decent employment; I don't want to talk about water and housing ... we have had to struggle for every bit of good thing we have ... now they are telling us that climate change is going to cause more floods and droughts which means even the weather is being turned against us. Then you ask me about paper rights ... they mean nothing if you have no potatoes or greens growing in your garden ... you are as good as dead.

When we asked Ndumiso Sondezi, a young community organiser from Inanda township in the north of Durban, what he thinks about the official COP17 negotiation process taking place at the ICC, he put it this way:

Those people gathered in there don't know how to reverse climate change because to reverse climate change means to let go of this modern lifestyle, reliance on petrol and burning coal for generating electricity.

Ndumiso is an unemployed youth, he has been a head of his household of four siblings since he was 15, when both his parents passed away due to illness. He is a community activist who is involved in many voluntary projects in Inanda and KwaMashu townships of Durban. He is of the view that the solution to climate change lies in radically altering all that which is considered 'normal' in the urban modern society and looking to rural communities as examples of how to use energy resources efficiently.

Amabhunu amnyama ... asenzeliwari, amabhunu amnyama ... asbangela ukushisa asibangela iwari'
Amabhunu amnyama ... asenzeliwari, amabhunu amnyama ... asbangela ukushisa asibangela iwari'

The above is an isiZulu protest song sung by activists while protesting outside the Democratic Socialist Movement event at the C17 People's Space venue at the University of KwaZulu-Natal (UKZN); activists staged a disruptive walk-out because of the bad conditions at the climate refugee camp, exhaustion from excessively long bus and train rides, hunger and no space to take direct action against the Durban COP17. Loosely translated, the song means: The Black Boers are causing us worry ... the Black Boers are causing us to burn up. It is a word play on the Boer; the previous oppressors and controllers of state power and its attendant violence, who have now been joined by aspirant members of the new black ruling class after transition to democracy in South Africa.

Hlengiwe Mkhize, a mother of three and four grandchildren from the rural areas outside of Pietermaritzburg, gave this response:

I have come as a mother from the rural areas to the march in order to add my voice to the people who are demanding attention from the government and NGOs to consider people living in the rural areas, and that climate change is really putting our lives in danger. I don't believe that these negotiations are legitimate because it is the same people causing pollution negotiating with their government friends about the direction of our lives (rural communities). Even the organisers of this march are all well fed city people who have no idea about what they are talking about only that it pays their salaries.

Hlengiwe informed us that she had to borrow money from her neighbours in order to attend the march because she was under the impression that they would make their way into the main conference hall at the International Convention Centre (ICC) and interrupt proceedings for the day. In which case, she had hoped to tell the COP17 delegates of the hardships of having no food because the rains have failed and seeing all your children leaving the rural areas to seek work in the cities, only to return sick or dying.

Mzonke Poni, a long-time community organiser in the Anti-Eviction Campaign based in Mandela Park, Western Cape, had this to say regarding the civil society process towards the COP17 activities in Durban:

I must say I have a whole lot of uneasy feelings regarding the whole manner in which things were approached. Although we can see a lot of effort went into the logistics, we commend this, but of course as grassroots organisers, we cannot but wonder and ask: who really is controlling the civil society space, who is coordinating the space or how the space is being managed? So far what a lot of us have seen is that this whole process has been a power relationship in terms of those that coordinate and also those that are excluded from the coordination. Whenever a space is created, within civil society, we must realise that when we talk about civil society then we are talking about different groups and these groups operate differently. Some are professionalised NGOs, you look at trade unions you know ... they are professionalised, you look at community-based organisations which are grounded at a community level. But when we come to events such as the COP, the professionals dominate the civil society voice and dictate the means to engage with the powers that be. There are all these divisions, in-terms of the use of language.

Having been involved in earlier social movement mobilisations against UN conferences, first in 2001 at the World Conference Against Racism (WCAR) held in Durban, and then in 2002 at the World Summit on Sustainable Development (WSSD), Mzonke bemoaned the apparent decline in radical politics in the social movement sector, noting that the spirit and rapport of the C17 process lacked the grassroots organising and movement building ethos that had characterised earlier mobilisations. Because of the concentration of discussions amongst a chosen few leaders, many activists were not well informed about the political strategy of civil society engagement with the official COP17. The reluctance by many activists who did make it to Durban to participate in C17 activities at the People's Space and a refusal to be made to play along in domesticated top-down political performances of the C17 aspirant elite seemed more of an issue than the lack of knowledge about the impact of climate change on poor people's lives.

> From Cape to Cairo (Azania, Azania, Azania)
> Morocco to Madagascar (Azania, Azania, Azania)
> iAzania yizwelethu, solilwela nge bhazooka ... (Azania, Azania, Azania)
> *[From Cape Town to Cairo, from Morocco to Madagascar ... Azania is our land we will fight for its liberation with our bazookas ...]*

The song above was sung during the spontaneous mass walk-out of social movement activists during the first plenary session of the Democratic Left Front (DLF) in the main lecture theatre venue of C17 People's Space; this led to a subsequent slow 'boycott' of many other activities in the People's Space by community activists. Those involved in the interaction lyrically registered protest or disagreement by pounding the floors and furnishings in a repertoire of supportive rhythms typical of the toyi-toyi genre of political refusal in song or musical arrangement. This, along with other militant songs sung by activists outside official venues and at the climate refugee camp, suggests that on the street the mood was less reconciliatory towards the activities at COP17. Despite intimidation and the exclusion of activists, they continued to speak/sing a language of contentious collective actions, bemoaning the domestication of South African activist response by C17 as a kind of NGO occupation of social movement space.

An old Mkhonto we Sizwe liberation isiZulu song was 'remixed' in the following manner:

> *Makuliwe... makuliwe omakuliwe makuliwe ngoba uObama akafuni ukusayina phansi makuliwe ... uyabhaleke nang'yabaleka uObama nezinja zakhe ...*
> [Let there be war because Obama does not want to commit and sign an agreement. Look Obama is running away, he is running away with his dogs.]

By the terms of these lyrics, the COP17 is framed as an occasion for confrontation equivalent to the days of apartheid. In the past, the place where the song names (United States (US) president Barack) Obama used to refer to an apartheid president, specifically HF Verwoerd, PW Botha or JG Strijdom. Now the symbolic leader of global apartheid is the US president and the rest of the countries of the global North who refuse to commit real emissions cuts and move to a carbon free economy. The point being that though activists were not given the space to articulate a clear critique of the COP17 within the official civil society process, the consciousness among most was as clear as daylight and those we interacted with during the course of the research felt at best disappointed by C17's 'sell-out' politics.

Bertha Swanerpool a soft-spoken activist from the Northern Cape had this to say:

> The pamphlets are saying climate change in also about gender justice, how are the C17 leaders showing sensitivity to gender issues by making women travel over 24 hours and then command them to get ready to march upon arrival? We came here because we have a real problem with our place, it is very historical, and in around 1816 after the wars at Witrivier, the colonial queen Victoria of Britain, gave land to the people. Five Xhosa families and some coloured families. All the children attend the same school. We are brown people, not only black; we work together, there is no difference on race. But for the past five years we have been experiencing more droughts and constantly run out of food we want to talk to other communities experiencing these difficulties because maybe together we can come up with solutions; we have not been able to do this because right after the march we are told to gather our things for the trip back home.

Looking visibly exhausted, Bertha was sitting on the side of the road leading up the Durban beachfront, with what looked like a bag of her clothes, which she took with her on the long march because of her concerns about the safety of her personal belongings at the refugee tent. She spoke in a regretful tone at having travelled all the way to Durban and the prospect of the long trip back to the Northern Cape.

Activists who came with the Africa climate caravan were also disappointed, we spoke to Violet Mncube, a social worker and rural women's rights activist from Zimbabwe:

> ... little has been said about the links between climate change and the prominence of lifestyle deceases like the increase in cases of HIV and AIDS, we need to be more vocal in criticising the role of industrialised countries especially their contribution to emissions of greenhouse gases. A lot has been said about how poor people and countries must adapt to climate change, yet not much has been learned from traditional and indigenous cultivation techniques like the case of Zimbabwe: we are using uncomplicated methods like discouraging the use of large-scale tractors for cultivation. Instead, people concentrate on digging single holes for cultivation of a combination of food crops that are mostly to meet the needs of a household, not commercial goals of profit making. We also strongly discourage the chemical fertiliser industry.

Violet felt she expressed the opinions of many activists from across the continent who had made the long journey to Durban. She shared how many activists expected much more critique of a Western capitalist development trajectory and a stronger Pan-Africanist solidarity because of the vulnerability of the African continent and its people due to climate change; especially since COP17 was taking place on African soil. Yet, in the language of C17, no such critique was articulated. This, for her, represented a compromise that did not capitalise on converging African voices around a common threat.

During the march on the Global Day of Action, a large group of people (about 400) dressed in green official 'COP 17 volunteer' tracksuit uniforms formed a rearguard of the march and began to physically attack other protesters who were chanting anti-capitalist and anti-African National Congress (ANC) songs or displaying placards critical of the policies of the South African government as they marched.

We interviewed Xolani Zungu one of the people in the 'Cop-17 volunteers' group who had this to say about COP17:

> South Africa civil society should be celebrating COP17 as another 'World Cup' event and proof that the South African president Mr Jacob Zuma is well loved by the international community … Mr Zuma is going to solve the climate change if people just vote for him … the purpose of the march was to go and congratulate and encourage the South African government on a job well done.

The above quote demonstrates the often-contradictory sentiments that prevailed amongst the marchers, who were all unified by the desire to demonstrate a show of strength for forces in South African civil society. The banner at the front of the march also added to this lack of clarity, as it read something like: 'Civil society united against climate change', which was a slogan not too different from the one used to brand the official Durban COP17 event. As a result, there were many informants who thought the march was a celebration of the South African government's success as the host of the global negotiations.

Thembekile Khumalo, street trader (selling socks, sweets, underwear and other items) at the Durban Warwick Market Junction area, pointed out that:

> My son, this conference business, whether it is about war or jobs whatever … you just have to look at who are the people speaking and what are their own conditions of life. We poor people need not be told about climate change and its dangers … we are already feeling them in our makeshift shacks or crumbling 'RDPs' [government-built housing scheme]; we don't have water not to talk about toilets, I have been unemployed for ten years as you see me sitting here selling these small items … When it starts raining; I become worried because I don't know if I will find my children and shack standing when I return home … the weather and the world has changed for the worse … especially for people like me, I don't need a conference to tell me that, I live it every day …

III MAKING SENSE OF THE TESTIMONIES

After extensive interactions with civil society actors at COP17, a picture that emerged was that politically, the march on the streets of Durban during the Global Day of Action could be read as a metaphor for the state of formal civil society in South Africa today. With the NGO leaders standing on raised platforms, having been seduced by the allure of the gleam and false comforts (bottled water, laptops and riding on truck transport as opposed to slogging it out on the streets) as signals of being liked and/or accepted by fat-cat politicians, remaining aloof from the monochromic T-shirted masses. At the back-end of this procession was a quasi-military formation – the 'Green Bombers', policing what are acceptable utterances and what are not, by threat of physical violence, the bodies of subalterns only useful for symbolic display to an external gaze of national and international media. The march became a poignant summary of all that has gone wrong in co-opted civil society spaces, it suggests that the form of articulation that formal civil society in South Africa is taking has become less imaginative and less progressively militant, a tamed civil society, which has become a rehearsed spectacle, acting more as an accomplice to power rather than a challenge to the established hierarchies of power.

At global forums such as typified by COP17, the views of ordinary people are often dismissed as having a shallow understanding of the complex issues being dealt with. Yet, micro perspectives such as those sampled in the quotations in part II above, can offer a powerful accompaniment to the silences and gaps in official archives; especially when an event is over and is in the process of being forgotten.[4] Using reported speech has presented an opportunity to interrupt the flow of history-making narratives around mega conferences such as the COP, and to give space to African marginalised voices to 'speak' through creative texts,[5] such as Qwasha! *Climate justice community dialogues compilation* vol 1. Our focus was less about ordinary people's agency in climate change debates as circumscribed in national media or represented in the formal COP17 literature (although this is very important), and it was more about taking respondents' experiential accounts about climate change, the COP and the alternative spaces opened for critical engagement and how these can link the past and present to the future of climate justice activism.[6]

Though framed in a repertoire of experiential life-histories, casual conversational expressions and in protest songs; most respondents we spoke to revealed an understanding of climate change as the rise in global temperatures and the failure of seasonal rain patterns resulting, for some in droughts, and for others in flooding. There was a shared sense that these changes had something to do with pollutants and human actions by large industries. In the context of South Africa's recent history of injustice, which saw the majority of the African population systematically expropriated from their arable lands and removing the possibility for self-sufficiency for food production. The unprecedented hikes in food prices are not an academic matter, as Emily Dlomo pointed out above.

The official COP17 talks focused on efforts to ultimately secure a global agreement to reduce the amount of CO_2 emissions (to levels of 350 ppm, according to mainstream scientific consensus, a minimum required to ensure that global temperature rise does not exceed an average 2°C increase in temperature levels); these attempts remain meaningless unless they also recognise that current levels of CO_2 are a result of the fact that countries in the global North have been industrialising since the Industrial Revolution, and that, this was done largely at the expense of the rest of the planet, especially people of the global South. As such, there is a debt owed, especially, to the unborn children of the global South. Ignoring this acknowledgement and the continued postponement of reaching an equitable agreement that can be globally enforced, ultimately translates to the perpetuation of contraventions of the rights to life, a healthy environment, adequate and nutritious food,

4   M Wright *Strategies of Slaves and Women: Life-Stories from East/Cental Africa* (1993) 12.
5   M Vaughn 'Reported Speech and Other Kinds of Testimony' in L White, SF Miescher & DW Cohan (eds) (2001) *African Words, African Voices. Critical Practices in Oral History.*
6   S Geiger *'Women's Life Histories: Methods and Content'* (1986) 11 *Signs* 334.

water, decent housing, and other related rights against the majority of the populations of the South. In contrast, the official talk about climate change in exclusively technicist and rational scientific parlance with scant (if any) reference to socio-historical justice issues made the COP17 seem, to ordinary people, just another occasion for developed countries and wealthy members of governments and the 'Next Government Officials' (NGOs) of the South to eat, strike deals and then postpone getting binding commitments to making substantial changes to the 'business as usual' approach taken by most countries of the North.

We observed that many respondents in and around the various activities during COP17, articulated themselves in songs and dance performance, it is for this reason that we felt that in this research project, a song cannot be just a song, echoing African art historian Margaret Drewal:

> In Africa, [musical] performance is a primary site for the production of knowledge, where philosophy is enacted, and where multiple and often simultaneous discourses are employed ... Not only that, but performance is a means by which people reflect on their current conditions, define and/ or reinvent themselves and their social world, and either reinforce, resist, or subvert prevailing social orders.[7]

Perhaps then, activists are on point when they sing that climate change negotiations cannot be 'business as usual' any longer; contestation is required, because US president, Barrack Obama, the leader of what is believed to be the most powerful country on earth is 'running away' from the responsibility to commit to making a change in the development model utilised by his country, along with 'his dogs' – referring to other developed countries who have similarly shown reluctance to commit to radically changing how production and profit is accumulated in the global North. In a sense the songs are a commentary about the legitimacy of the entire COP negotiation process leading up to what was the 17th occasion in Durban. The implication is that until the day on which these climate negotiations move from recognition of justice and equity to being at the centre of all talk of change, only then may the legitimacy of the entire process be restored in the eyes of ordinary people. There is a basis for this view; one needs only consider the fact that, so far, none of the developed countries and mainstream media in general, has emphasised the human rights dimension and rights of the planet when discussing the climate change crisis. Instead, dominant narratives from the countries of the North emphasise that countries of the South are equally responsible to reducing emissions and adapting to the consequences of climate change.

In general, many of the respondents of the interviews and conversations felt that they were largely barred from most of the official COP17 venues, as well as being left out of the formal orchestrated civil society programme (though many participated in the march on the Global Day of Action). Most raised the critique of the climate change within a generalised critique of capitalism as a socio-economic system, many pointed to inequalities and the proliferation of injustice as a result of capitalism.[8] The crisis of climate change is not separated

---

7   MT Drewal *Yoruba Ritual: Performers, Play, and Agency* (1992) 17.
8   Wright (note 4 above).

from long-standing modes of crisis within which people on the margins are forced to exist.[9] Climate change and outside mobilisation of rights discourses are understood not in isolation from other pressing mobilisations such as those against privatisation of water or land evictions to make way for industrial agriculture and so on.

Thembekile Khumalo's frank sentiments seem to capture the tone which most of the respondents adopted in relation to COP17. Most were sure that there need not be a conference held 17 times in order to realise that something drastic has to be done about climate change and that capitalist production is the main culprit behind the industrial pollution and poisoning of the environment. As such, conferences represent the extension of business in that they end up self-perpetuating themselves, not for the sake of finding solutions, but rather by becoming *the* 'business' – to paraphrase Thembekile Khumalo. Some even make a career out of and constantly pursue networking opportunities from each successive meeting in a different part of the world. Violet Mncube pointed out that on the ground, poor people especially African women, have had to absorb the additional burdens of uncertainty and disease. From these women's perspective, then, COP17 was not meant for them since they need not be told by a conference that their quality of life has worsened, this is a reality they 'live (with) everyday' as Thembekile Khumalo put it. In a similar vein, Hlengiwe Mkhize felt that many governments and NGOs pretend to speak particularly for African rural women, when in actual fact little practical action is done to ameliorate the conditions of life for rural people; yet conference after conference makes reference to rural vulnerability.

Furthermore, the elusive average of 2°C within which global warming should be contained remains strongly contested by most African activist and other allied networks, arguing that it is still dangerously high, with a potential for destructive impacts on biodiversity, subsistence agriculture and regularity of extreme weather (excessive flooding and regular droughts), all of which threaten the quality of life of African populations.

Regarding the C17 civil society process, Mzonke Poni's 'uneasy feelings' are not without basis either; as we too observed from the onset that there remained large divisions on what a shared civil society position should be regarding the UNFCCC. The members of the C17 were adamant that in the interest of maintaining unity, the C17 would not engage the substantive issues up for negotiation at the UFCCC, nor, philosophical, ideological, political matters linked to the climate change negotiation processes; these would be left to individual organisations and C17 leaders citing that 'there are differences, with some organisations working with business and some being anti-capitalist, which meant it was difficult to do common messaging'.

However, a closer examination of the groups leading the C17 reveals that most have been members of the Climate Justice Now! Activist network, and as such, were part of a political and ideological critique of the COP negotiations; this then begs the question, why did these organisations adopt a neutral stance

9   L White *Magomero: Portrait of an African Village* (1987).

when they had been privy to the contestations that characterised global civil society's engagement with the COP process? The programming of much of the C17 activities was done to the exclusion of critical input from grassroots' activists to the extent that there was no space provided prior and during the COP17 event for discussions on political strategies, let alone political action beyond the carnival march. Therefore, it was not surprising to hear views of people like Xolane Zungu along with other 'Green Bombers', who believed that the march on the Global Day of Action was aimed at congratulating the South African government for a job well done.

While at the same time, people like Bertha Swanerpool expressed the opposite sentiments of complete lethargy and deep disappointment that so much organising energy had gone into putting on a show for high government officials rather than focusing on building relationships among the multitude of activists attempting to create strategies for harnessing a grassroots' movement for climate justice. In this way the C17 initiative is regarded as being short sighted, amounting to silencing the possibility of a unified civil society initiative to go beyond the COP17 event as the only frame of reference for articulating a politics of climate justice. Far from leaving activists empowered and energised, the activity left them feeling disorientated and discarded (in their terms: 'as used condoms') after playing a political game of some unseen committee (C17) situated in a phallic tower (UKZN) on an aloof hill looking down at the refugee camp where the makeshift tents for poorer activists were located.

IV BY WAY OF CONCLUSION

We went into the 'field' armed with clear questions, a reasonable amount of theoretical readings on contemporary activist issues and backgrounds of being community media activists ourselves with a modest research strategy that aimed at getting responses that could be checked against other sources for factuality and truth-worthiness. What we came out with was a rich collection of life stories where people told us what they wanted more often than what we wanted to hear. The subjects of the many conversations and interviews we held, spoke back at us in a way that no document could ever do – no matter how well written. Giving us a glimpse of the world according to how they saw it (or sang it). These glimpses, fragmented as they may be, provided the tools with which a different construction of the past-present could begin to take shape.

For us, the Qwasha! *Climate justice community dialogues compilation* vol 1 was first and foremost, an artistic contribution. After all, stories need not be either true or false and need not be proven beyond being talked about. They do, never the less, carry other types of empirical valorisation, like additional commentaries on apparently unrelated subjects that have a bearing on the truth-worthiness of the story being told at any point in time. Furthermore, our effort in Qwasha! *Climate justice community dialogues compilation* vol 1 was targeted at testing new ways of finding out how people at the margins create, express and remember their own selves. We were enriched by the effort and learnt lessons about the limitations of our effort. There is a need to properly ground community archival efforts in the actual communities themselves where genuine relationships can be nurtured and outputs shared as a means of deepening the bonds among members. As it were, the COP took us away from our communities into the conference circuit where we were constantly treated as outsiders.

Very limited resources meant that we stretched ourselves too thinly and exhausted the little resources available on an event that bore very little results for people on the ground. However, the experience of co-habitation, constant production meetings and field application has built lifelong bonds among collective members and was a necessary step towards a more long-term arrangement. Based on the experiences of producing Qwasha! *Climate justice community dialogues compilation* vol 1, we have begun a conversation on how best to re-frame and refine texts on the online archive in such a way as to allow them (the marginal texts and voices) to speak for themselves.

MONEY SPEAKS

WATER RIGHTS, COMMONS AND ADVOCACY NARRATIVES

P~~ATRICK~~ B~~OND~~

I I~~NTRODUCTION~~

How are human rights articulated within water advocacy movements, and what lessons – especially from South Africa – can be drawn from subsequent efforts to introduce rights into climate advocacy? Is rights language a strong countervailing force to the market's commercialising tendencies when it comes to water or climate policy? Or is it co-optable *within* neo-liberal environmentalism, thus requiring a different political framing?

This chapter reviews the limits to a typical 'rights-talk' legalistic counterstrategy favoured by liberal non-governmental organisations (NGOs). This rhetoric is usually contrasted with neo-liberal 'Green Economy' narratives that aim to price environmental services. The latter narratives dominated the Rio+20 Summit in June 2012 and are set to continue to affect socio-environmental policy in coming years. Along with 'Payment for Ecosystem Services', carbon trading and geo-engineering, one of the most disputed aspects of 'neo-liberalised nature' is water privatisation. The main framing device against privatisation adopted by a vast set of social movements and allied NGOs across the world since the late 1990s, is making the 'right to water' a central demand. In the pages below, I consider the case of water, drawing on detailed experiences with rights narratives and market pressures in Johannesburg, South Africa (which, coincidentally, hosted Rio+10 in 2002), and then consider how these lessons apply to the emerging Climate Justice movement's narratives.

In the process, I must ask what kind of analyses, strategies, tactics and demands worked best under circumstances of intensifying global struggle for socio-environmental justice? Specifically, this chapter asks, how appropriate is it for influential strategists to emphasise rights talk as a means of countering corporate-neo-liberal pressure, such as in the field of water? The global debate trails by a few years the same concerns about the efficacy of water rights invoked both by the state and by social activists in post-apartheid South Africa. The simple question is whether by invoking the right to water and attempting to define it in the context of neo-liberal municipal management, does a generic commitment to rights trump the market? Or instead, does rights-talk work *within neo-liberalism*?

The critical requirement of the neo-liberal project when applied in micro-developmental settings is the imposition of market logic. What that means in the case of retail water and sanitation is to achieve 100 per cent cost-recovery on every drop sold, and where there are subsidies (and especially cross-subsidies), to remove these because they represent inefficiencies, and

147

deterrents to market investments. Such a logic can be readily applied not only by a private supplier seeking a cost+markup to achieve a typically desired rate of 30 per cent on investment in the water sector, but also by municipalities whose water departments are increasingly run as independent cost-centres. It is in this sense that the drive to privatise water can occur within a municipality, even with full public ownership. The point is that the logic of the market – the neo-liberal opposition to subsidisations and other price distortions – is the opposite of the logic of society and environment. The question is whether the rights discourse offers a sufficient basis for framing opposition to market logic.

To illustrate the conflict, Durban provides the best data to judge the efficacy of using pricing measures as a mechanism of achieving basic water rights at the same time as demand management (conservation) while still operating within the 100 per cent cost-recovery logic. Research conducted at the University of KwaZulu-Natal (UKZN) by Chris Buckley and former city official Reg Bailey showed that water 'price elasticity' – the negative impact of a price increase on consumption – for the city's highest-income third of the population is 0.10. A doubling of the real (after-inflation) water price from 1997 to 2004 generated less than a ten per cent reduction in use. (What was proposed by Johannesburg for high-volume users was not a 100 per cent real increase, but a meagre three per cent rise – ten per cent in nominal termsbut inflation was seven per cent.) Durban research revealed that instead, the impact of higher prices is mainly felt by low-income people, who recorded a much larger 0.55 price elasticity.[1]

Likewise, international studies suggest that while levels of water consumption may dip following large price increases, patterns of use generally reassert themselves fairly quickly in all but the lowest income groups.[2] Ironically, as the 'right to water' was fulfilled through an official commitment to Free Basic Water, as explored below, *the result of price changes at higher blocks in Durban and Johannesburg was further water deprivation for the poor alongside increasing consumption in the wealthier suburbs*, which is in turn creating demand for more bulk water supply projects – including another extremely expensive Lesotho Highlands Water Project dam – which will then have to be paid for by all consumers, and which will have major environmental impacts.

To track the prospects of the rights and commons framing, I first consider some crucial background contextual information about South Africa, including the challenge of water/sanitation delivery, followed by consideration of the specific problems associated with Africa's richest city (Johannesburg), particularly its most politicised neighbourhood (Soweto). The chapter then addresses thorny technical issues that have arisen in the course of transforming rights discourse into justiciable service delivery. The limits of liberal capitalist democracy as the basis for social services provision in poor neighbourhoods – under circumstances of extreme inequality and fiscal

---

1   R Bailey & C Buckley 'Modelling Domestic Water Tariffs' presentation to UKZN CSS (7 November 2005).
2   V Strang *The Meaning of Water* (2004).

pressures – became evident in 2009, when Soweto activists promoting water rights were defeated in the courts. Their potential move 'out of the box' of the liberal rights narrative, towards a 'commons' approach to water, is explored in the Conclusion, where the next terrain of crucial socio-environmental struggle is being joined: the climate.

II   Is SOUTH AFRICA RIGHTING WATER WRONGS, OR MERELY REVISITING
      RETAIL RIGHTS?

Since the United Nations (UN) Declaration of Human Rights, the idea that all individuals have certain basic human rights, or entitlements to political, social, or economic goods (such as food, water, etc) has become a key framework for politics and political discourse. In appealing to human rights, groups and individuals attempt to legitimise their cause, and to accuse their opponents of 'denial of rights'. As water is essential to human life, social conflict surrounding water is now framed in terms of the 'human right' to water. In this 'culture of rights', social groups use 'rights talk' as a blanket justification for the provision of water; in some cases, however, even popularly elected governments dispute their exact responsibilities for water provision and management.

During apartheid, water was a relatively low-cost luxury for white South Africans, with per capita enjoyment of home swimming pools at amongst the world's highest levels. In contrast, black South Africans largely suffered vulnerability in urban townships and in the segregated 'Bantustan' system of rural homelands, which supplied male migrant workers to the white-owned mines, factories and plantations. These rural homelands had weak or non-existent water and irrigation infrastructures, as the apartheid government directed investment to the white-dominated cities and suburbs, and also in much more limited volumes to black urban townships.

After 1994, racial apartheid ended, but South Africa immediately confronted international trends endorsing municipal cost-recovery, commercialisation (in which state agencies converted water into a commodity that must be purchased at the cost of production), and even the prospect of long-term municipal water management contracts roughly equivalent to privatisation. At the same time, across the world, commercialisation of water was being introduced so as to address classic problems associated with state control: inefficiencies, excessive administrative centralisation, lack of competition, unaccounted-for-consumption, weak billing and political interference. Across a broad spectrum, the commercialisation options have included private outsourcing and the management or partial/full ownership of the service. At least seven institutional steps that can be taken towards privatisation: short-term service contracts, short/medium-term management contracts, medium/long-term leases (affermages), long-term concessions, long-term Build (Own) Operate Transfer contracts, full permanent divestiture, and an additional category of community provision which also exists in some settings. Aside from French and British water corporations, the most aggressive promoters of these strategies are a few giant aid agencies, especially USAID, the British

Department for International Development, and the World Bank. As a result of pressure to commercialise, water was soon priced beyond the reach of many poor South African households, resulting in an estimated 1.5-million people disconnected each year due to inability to pay by 2003.[3]

The Constitution of the Republic of South Africa, 1996 however, included socio-economic clauses meant to do away with the injustices of apartheid, including, 'Everyone has the right to have access to sufficient food and water' and 'Everyone has the right to an environment that is not harmful to their health or well-being'.[4] The Water Services Act 108 of 1997 put these sentiments into law as 'the main object': 'the right of access to basic water supply and the right to basic sanitation necessary to secure sufficient water and an environment not harmful to human health or well-being'.[5] Grassroots water activists seized on these guarantees to clean water and their discourses soon invoked rights talk. They insisted upon a social entitlement to an acceptable supply of clean water, amounting to at least 50 litres supplied per person per day, delivered via a metering system based on credit, not 'pre-payment'.

The surge in confidence for the rights narrative left their critics bemoaning a new 'culture of entitlement' in which the government was expected to solve all social ills. As Lungile Madywabe of the (pro-market) Helen Suzman Foundation put it:

> Cynics fear that a culture of entitlement is growing. But the left finds such statements insulting and dehumanising, and argues that it is crass to suggest that people are unwilling to pay for services when unemployment exceeds 40 per cent ... A turning point in the African National Congress government's thinking came in 1995, when Nelson Mandela returned from Europe and spoke in favour of privatisation.[6]

The commercialisation of water was viewed with great enthusiasm by the new South African government. In South Africa, the shift to a market-based system of water access has been protested in various ways, including informal/illegal reconnections to official water supplies, destruction of prepayment meters, and even a constitutional challenge over water services in Soweto. While such protests confront powerful commercial interests, they attempt to shift policy from market-based approaches to those more conducive to 'social justice'. Nevertheless, this chapter draws on the 2008 to 2009 courtroom dramas to argue that a rights discourse has significant limitations so long as it remains primarily focused on the social domain.

The objective of those promoting water rights should be to make water primarily an eco-social, rather than a commercial, good. Including eco-systemic processes in discussions of water rights potentially links consumption processes (including over-consumption by firms and wealthy households) to environmental sustainability. However, the lawyers developing strategy in the seminal case I consider below decided to maintain only the narrowest

---

3   M Muller 'Turning on the Taps' *Mail & Guardian* (24 June 2004).
4   Constitution s 27(1)*(b)*.
5   Section 2*(a)*.
6   L Madywabe 'A Compelling Need for African Innovation' The Helen Suzman Foundation (2 March 2005) 1 <http://70.84.171.10/~etools/newsbrief/2005/news0303.txt>.

perspective of household water usage, since to link with other issues would have complicated the simple requests for relief. Hence, given the lawyers' defeat, once I interrogate the limits to rights discourse in the South African context, the most fruitful strategic approach may be to move beyond the 'rights' of consumption to reinstate a notion of 'the commons', which includes the broader hydropolitical systems in which water extraction, production, distribution, financing, consumption and disposal occurs.

III WATER RIGHTS AND WATER DENIAL IN SOWETO AND JOHANNESBURG

One of the critical disputes in Johannesburg during the period 2001 to 2009 was interpretation of the African National Congress's (ANC) promise of a universal free basic water service. In the 2000 municipal election campaign, the ruling party's statement had been clear: 'The ANC-led local government will provide all residents with a free basic amount of water, electricity and other municipal services so as to help the poor. Those who use more than the basic amounts, will pay for the extra they use.'

There is an extensive record regarding the way the right to water was distorted in Johannesburg.[7] Initially, in 2001 Johannesburg Water officials reinterpreted the 'right to water' mandate regressively by adopting a relatively steep-rising tariff curve. In this fee structure, all households received 6,000 litres per month for free, but were then faced with a much higher second block (ie the curve was convex-up), in contrast to a concave-up curve starting with a larger lifeline block, which would have better served the interests of lower-income residents. The dramatic increase in their per-unit charges in the second block meant that for many poor people, there was no meaningful difference to their average monthly bills even after the first free 6kl. Moreover, the marginal tariff for industrial/commercial users of water, while higher than residential, actually declined after large-volume consumption was reached.

In early 2008, changes to Johannesburg Water pricing policy meant that although there was a higher Free Basic Water allotment, of 10kl/month, the 2000 promise of free basic water would be kept only for the small proportion of the population declared 'indigent', instead of on a *universal* basis to 'all' residents. Facing the lawsuit by Soweto residents and their high-profile lawyers, and following the departure of the French water company that set the original prices in 2006, there was scope for a slightly more redistributive and conservationist pricing system, and the 2008/09 water price increases included very slight above-inflation rises for higher blocks of consumption.

What ideology informed Johannesburg officials' orientation to water pricing? Even though the municipal officials insisted that they were meeting the basic rights obligations in the Constitution and prevailing water law, the top-down neo-liberal approach to meters and consultation conformed to the city's overall strategy of decentralisation and geographical differentiation of service provision according to ability to pay. The World Bank reported on its:

7 For more background on South Africa's water conditions discussed below, see, P Bond *Unsustainable South Africa* (2002); and P Bond *Talk Left Walk Right* (2006).

local economic development methodology developed for the City of Johannesburg in 1999. The latter sought to conceptualize an optimal role for a fiscally decentralized City in the form of a regulator that would seek to alleviate poverty by applying a two-pronged strategy. The first prong would focus on reducing 'income-poverty' through job creation by creating an enabling business environment for private sector investment and economic growth in Johannesburg. The second prong would address non-income poverty reduction by directly tracking the effects of local government expenditures on service delivery to poor households in the city.[8]

The 'enabling business environment' kept prices low for business but high for the poor, notwithstanding the 'second prong'. Moreover, the Bank encouraged the commercialisation of the municipal water company, which led to one of the world's largest management contracts, won by the French firm Suez for the period 2001 to 2006. As the world's second largest water company, Suez came to South Africa just before the end of apartheid, picking up three small water concessions in Eastern Cape towns during the early 1990s. The firm won the bid for a five-year trial contract to manage Johannesburg Water, in part by taking the city's councillors on a junket to Argentina the year before, where the 'success story' of Buenos Aires was unveiled.

The Suez contract in Buenos Aires would fail when the Argentine government disallowed Suez's substantial hard-currency profit repatriation in the midst of the 2002 economic crisis.[9] Yet in adapting to the challenge posed by social movement rights activists, Suez came to realise that it could 'strongly' endorse water and sanitation as a 'fundamental right', so as to acquire more business opportunities:

> through its partnerships with local authorities, by working in the southern hemisphere to provide an additional 11.8 million people with access to drinking water [and] an additional 5.7 million people with access to sanitation.[10]

The reality, however, was that when confronted with a great many poor people, Suez resorted to water self-disconnections and regressive pricing policies, as Soweto residents soon experienced.

In South Africa, Suez inherited a dysfunctional retail water system, especially in Johannesburg's vast shack settlements, which are home to nearly a third of the city's 3.2-million residents. There, according to city surveys at the time Suez entered, 65 per cent of the population use communal standpipes and 20 per cent receive small amounts of water from tankers (the other 15 per cent have outdoor yard taps). For sanitation, 52 per cent have dug pit latrines themselves, 45 per cent rely on chemical toilets, two per cent have communal flush toilets and one per cent use ablution blocks. These conditions are particularly hostile to vulnerable people: they breed opportunistic infections

---

8   World Bank 'South Africa: Monitoring Service Delivery in Johannesburg' (2002).

9   Also in 2002 the Lesotho government prosecutors charged Suez subsidiary Dumez with bribing Masupha Sole, the manager of the Lesotho Highlands Water Authority (which supplies Johannesburg with water). Sole allegedly received US$20,000 at a Paris meeting in 1991 to engineer a contract renegotiation providing Dumez with additional profits in excess of US$1-million, at the expense of Johannesburg water consumers. On those grounds, the South African Municipal Workers Union (SAMWU) asked Johannesburg officials to bar Suez from tendering for the water management contract, but this request was refused.

10 Suez 'Promoting Access to Water and Sanitation' <http://www.suez-environnement.com/water/challenges/promoting-access-water-sanitation/>.

at a time when Johannesburg's HIV rate has soared above 25 per cent, and in the last decade cholera and diarrhoea epidemics have killed many tens of thousands of people, especially children.

But instead of expanding water access in these underserved areas, Suez initiated massive water disconnections. In early 2002, just before community resistance became an effective countervailing force, Johannesburg officials were disconnecting more than 20,000 households per month from power and water, contradicting the claim on the Department of Water Affairs and Forestry's website that Johannesburg offers 100 per cent of its residents Free Basic Water. For municipal bureaucrats and Suez, disconnecting low-income people and maintaining low water/sanitation standards was a strategy, quite simply, to save money.

Suez began its management of Johannesburg's water by installing 6,500 pit latrines, a pilot 'shallow sanitation' system and thousands more pre-payment water meters in poor areas, including Soweto. Pit latrines require no water. The shallow sewage system was only attempted sporadically due to consumer dissatisfaction. With this system, maintenance costs are transferred to so-called 'condominium' residential users, where a very small water flush and slight gravity mean that the pipes must be manually unclogged every three months (or more frequently) by the residents (typically women) themselves. The water-borne system breaks down, thus, not by accident – but *by design*, so as to save the city water.

As for the payment system, unlike conventional meters in wealthy suburbs, which provide due warning of future disconnection (and an opportunity to make representation), pre-payment meter disconnection occurs automatically and without warning following the exhaustion of the 6,000 litre free water supply. If the disconnection occurs during the night or over a weekend when water credit vendors are closed, the household has to go without water until the shops are open again, and if the household does not have money for additional water, it must borrow either money or water from neighbours in order to survive. The *Mazibuko* plaintiffs argued that the pre-payment water meter represented not only a threat to dignity and health, but also a direct risk to life in the event of a fire. Dangers from inadequate water resulting from self-disconnecting pre-payment meters were starkly illustrated when two children died in a shack fire in 2002, which in turn catalysed the lawsuit by five Sowetans (four of whom were women) that became known as *Mazibuko* v *Johannesburg Water*, or the 'Phiri' case after the area of Soweto where the plaintiffs resided.

One central problem was that Johannesburg managers were reluctant to offer a rising block tariff so as to redistribute water from rich to poor. If designed properly such systems penalise luxury consumption and promote conservation. In 1996, this potential was demonstrated in the Hermanus municipality, which raised prices on high consumption through a steep block tariff. Within four months, per capita peak demand for bulk water was reduced by one-third, while revenues increased by one-fifth.[11] In Johannesburg, in contrast, the block tariff adopted in 2001 was highly convex so that the additional marginal price increases for wealthier, high-volume users were negligible.

The block tariff system applied in Johannesburg reflected Suez's logical opposition to water conservation, for its self-interest is selling more water to those people who could pay for it. The increasingly expensive water Suez supplied to Johannesburg was piped hundreds of miles across the Lesotho mountains in Africa's largest cross-catchment water transfer, which caused a five-fold increase in water prices, from US$0.30 to US$1.30/kl during the late 1990s. As Johannesburg water customers became liable for Lesotho dam loan repayments, they faced an average 69 per cent increase in the nominal cost of water supply from 1996 to 1999, with high-volume users paying a much lower increase. By the time the city's commercialisation strategy was established in 1999, Johannesburg's water prices had become more regressive than even during the apartheid era (ie with a flatter slope in the block tariff).

In sum, rights advocates argued, the underlying problem was that across South Africa, the self-interest of powerful municipal constituents – large businesses, farms and rich ratepayers – was to keep water prices relatively low, which in turn required limiting provision in low-income neighbourhoods. In this context, rights advocates accused the city of adopting the following strategies: (1) imposition of water prices that soar after a very small, free amount of roughly two toilet flushes per person/day for member households, so that the next block of consumption becomes unaffordable; (2) disconnection of people too poor to pay for any water beyond the 6kl allocation; (3) offering Free Basic Water on the basis of a *household* as a unit (rather than the ANC's 1994 Reconstruction and Development Programme (RDP) recommendation of 50 litres per *person* per day), which penalised larger families and those who have backyard shackdwellers or tenants who also drew upon the per-household supply; (4) provision of low-quality water and sanitation technology to tens of thousands of poor households, with the objective of reducing consumption (the technology includes pre-payment water meters, chemical toilets, Ventilated Improved Pit Latrines, and 'shallow sewage' systems featuring smaller pipes and lower gradients, no cistern for flushing, and manual unclogging of faeces when pipes periodically clog); and (5) provision of differential technology according to geography, race and class, such that water-saving hardware was only imposed on people in townships and informal settlements and not in wealthier and whiter suburbs.

In March 2008, the water rights activists complained about three new Johannesburg Council innovations: (1) use of an inaccurate register of indigency, one that recorded only a small proportion of the city's poor and thus excluded a large number of low-income people from free water allocations; (2) a new system of 'means testing', even though gaining indigency status

---

11   S Wolfe 'Reforming and Rebuilding: Water Efficiency Initiatives in Hermanus, South Africa' *Water Efficiency* (2007) <http://www.waterefficiency.net/november-december-2007/american-water-purveyors-2.aspx>.

initially entailed an invasive process of surveillance; and (3) termination of the policy of universal free water services for all, even though termination directly contradicted the Constitution, the RDP and the ANC municipal election promise that 'all residents' would receive free services.

Resistance strategies and tactics developed over time. Initially, activists took what was already a popular township survival tactic – illicitly reconnecting power once it was disconnected by state officials due to non-payment (in 2001, 13 per cent of Gauteng's connections were illegal) – and added a socialist, self-empowered ideological orientation. Within a few months of Johannesburg Water's official commercialisation in 2000, the Anti-Privatisation Forum (APF) was formed to unite nearly two-dozen community groups across Gauteng, sponsoring periodic mass marches of workers and residents. The network also shared information with water activists across the world, for example in Cochabamba, Bolivia, Argentina, Accra, and Detroit. And from the APF came the Coalition Against Water Privatisation (CAWP), which assisted five of Soweto's Phiri neighbourhood activists to launch the Constitutional Court case in 2004.

Suez's water management in Johannesburg generated not only social conflict but also strife within the council, and the company's contract was not renewed in 2006, in spite of the desired 25-year extension option available in the original water commercialisation Business Plan. That plan had anticipated that (after-tax) profits from Johannesburg water supply would soar from R3.5-million (roughly US$300,000) in 2000/2001 to R419-million (US$50-million) in 2008/2009. One reason for Suez's departure was that Johannesburg Water's tactics were so hotly contested by the rights advocates, who had expected the Bill of Rights socio-economic clauses to be enacted.

In October 2009 South Africa's Constitutional Court overturned a seminal finding in lower courts that human rights activists had hoped would substantially expand water access to poor people. In the first case in the Johannesburg High Court, five Soweto women had successfully argued for their right to a larger supply of free municipal water and for abolishing the recently-installed pre-payment meter system. In the ruling, Johannesburg High Court Judge Moroa Tsoka ruled that the 'prepayment water system in Phiri Township' was 'unconstitutional and unlawful', and ordered the city to provide each applicant and other residents with a 'free basic water supply of 50 litres per person per day and the option of a metered supply installed at the cost of the City of Johannesburg'.[12] Judge Tsoka accused city officials of racism for imposing credit control via prepayment 'in the historically poor black areas and not the historically rich white areas'. He noted that meter installation apparently occurred 'in terms of colour or geographical area', and the community consultation process was 'a publicity stunt' characterised by

---

12   In the High Court: *Mazibuko v City of Johannesburg* 2008 (4) All SA 471 (W); in the Supreme Court of Appeals: *City of Johannesburg v Mazibuko* 2009 (3) SA 592 (SCA); 2009 (8) BCLR 791 (SCA); 2009 (3) All SA 201 (SCA); in the Constitutional Court: *Mazibuko v City of Johannesburg* 2010 (4) SA 1 (CC); 2010 (3) BCLR 239 (CC).

a 'big brother approach'.[13] It was the first South African case to adjudicate the constitutional right of access to sufficient water[14] as a matter of public (municipal) policy.

The hope from the April 2008 High Court ruling was that Tsoka had begun a new era of ecological, rational and more egalitarian water provision. However, 11 months later, the Supreme Court judgment ordered, whimsically, a decline in free water available per person from 50 litres each day to 42, *if the consumer can prove household 'indigency'*. The Supreme Court also found that prepayment meters were illegal according to Johannesburg Water's own water policy, but that the city didn't have to remove its illegal meters in Phiri, and instead could 'legalise the use of prepayment meters' by changing policies on disconnections to permit them without any administrative-justice process.

On the first point, the CAWP argued that 42 litres per person per day:

> falls short of what is universally accepted and recognised as the minimum amount of water needed for basic human needs and dignity. Even more problematic though, is that the Supreme Court's order to the City to provide this amount, is conditional. The very same City that has, at every opportunity, resisted the legitimate claims and demands of poor communities for adequate amounts of free basic water, is effectively allowed carte blanche (through its own assessment of what constitutes 'reasonableness' and 'through available resources') to determine the timing, character and extent of changes to its existing 'free water policy'.[15]

The Centre for Applied Legal Studies (CALS), which represented the applicants throughout the High Court case and beyond, agreed, 'The relief granted by the Court is neither appropriate nor effective ... [and] fails to address the City's constitutional obligations to progressively realise the amount of water it provides'.[16] But neither the activists nor the lawyers were persuasive in the final test, the appeal of the Supreme Court's judgment to the Constitutional Court, which handed down a ruling completely vindicating Johannesburg Water in October 2009. The judgment confirmed the original 25 litres per person per day plus pre-payment meters as 'reasonable and lawful'.[17]

The CAWP was infuriated, charging the Court with 'a lazy legalism and wholly biased and contradictory reasoning ... It is as if the thousands of pages of evidence and testimony provided by the Phiri applicants in countering the same from Johannesburg is simply ignored and/or considered irrelevant'. The CAWP was especially annoyed that the Court agreed Johannesburg had passed the 'progressive realisation' bar, interpreted by CAWP as allowing 'the state to do whatever it pleases, whenever it pleases and at whatever pace pleases it'. The CAWP also disputed the Court's definition of 'discontinuation':

> The water supply does not cease to exist when a pre-paid meter temporarily stops the supply of water. It is suspended until either the customer purchases further credit or the new month commences with

13   P Bond & J Dugard 'The Case of Johannesburg Water: What Really Happened at the Pre-paid "Parish Pump"?' (2008) 12 *Law, Democracy and Development* 1.

14   Constitution (note 4 above).

15   CAWP 'Phiri Water Case: Constitutional Court Fails the Poor and the Constitution' press statement (2 October 2009).

16   CALS press statement (2 October 2009).

17   *Mazibuko v City of Johannesburg* 2008 (4) All SA 471 (W).

a new monthly basic water supply whereupon the water supply recommences. It is better understood as a temporary suspension in supply, not a discontinuation.[18]

CAWP's reply: 'an insult both to the poor and to the constitutional imperatives of justice and equality'.[19]

IV   THE LIMITS OF THE RIGHTS NARRATIVE

Some argue that the whole basis of rights discourse (not just judgments like the Constitutional Court's in the Phiri case) exhibit the problems described above in part because of the rights movement's 'domestication' of the politics of need.[20] But more can be said about the intrinsic role of rights law from this standpoint, which allows us to question the legalistic reliance upon the rights narrative for popular access to water.

For example, Marius Pieterse argues that:

the transformative potential of rights is significantly thwarted by the fact that they are typically formulated, interpreted, and enforced by institutions that are embedded in the political, social, and economic status quo ... the social construction of phenomena such as 'rights' and 'the state' legitimize a collective experience of alienation (or suppression of a desire for connectedness) while simultaneously denying the fact of that experience.[21]

He provides a delightful illustration of this alienation – one that may well be felt by Phiri residents – in asking us to conceive of:

the South African socioeconomic rights narrative as a dialogue between society (as embodying the social and economic status quo) and certain of its members (a social movement, interest group, or individual seeking to assert herself against the collective of the status quo) over the satisfaction of a particular socioeconomic need. Behold, accordingly, the following three-act drama:

ACT 1: On the Streets
Member/Citizen: I am hungry.
State/Society: (*Silence*) ...
Member/Citizen: I want food!
State/Society: (*Dismissive*) You can't have any.
Member/Citizen: Why?
State/Society: You have no right to food.
Member/Citizen: (*After some reflection*) I want the right to food!
State/Society: That would be impossible. It will threaten the legitimacy of the constitutional order if we grant rights to social goods. Rights may only impose negative obligations upon us. We cannot trust courts to enforce a right to food due to their limited capacity, their lack of technical expertise, the separation of powers, the counter-majoritarian dilemma, the polycentric consequences of enforcing a positive right, blah blah blah ...
Member/Citizen: (*Louder*) I want the right to food!!
State/Society: (*After some reflection*) All right, if you insist. It is hereby declared that everyone has the right to have access to sufficient food and water and that the State must adopt reasonable measures, within its available resources, to progressively realize this right.
Member/Citizen: Yeah! I win, I win!
State/Society: Of course you do.

---

18   CAWP 'One Step Forward, Two Steps Back' press statement (2 October 2009).
19   Ibid.
20   T Madlingozi 'Good Victim, Bad Victim: Apartheid's Beneficiaries, Victims and the Struggle for Social Justice' in W le Roux & K van Marle (eds) *Law, Memory and the Legacy of Apartheid: Ten Years after AZAPO v President of South Africa* (2007) 107–26.
21   M Pieterse 'Eating Socioeconomic Rights: The Usefulness of Rights Talk in Alleviating Social Hardship Revisited' (2007) 29 *Human Rights Quarterly* 796.

ACT 2: In Court

Member/Citizen: I want food, your honor.

State/Society (Defendant): That would be impossible, your honor. We simply do not have the resources to feed her. There are many others who compete for the same social good and we cannot favor them above her. If you order us to feed her you are infringing the separation of powers by dictating to us what our priorities should be. We have the democratic mandate to determine the pace of socioeconomic upliftment, and currently our priorities lie elsewhere.

Member/Citizen: (*Triumphantly*) But I have the right to food!

State/Society (Court): Member/Citizen is right. It is hereby declared that the State has acted unreasonably by not taking adequately flexible and inclusive measures to ensure that everyone has access to sufficient food.

Member/Citizen: Yeah! I win, I win.

Everyone: Of course you do.

ACT 3: Back on the Streets

Member/Citizen: I am hungry.

State/Society: (*Silence*) ...

Member/Citizen: I want food!

State/Society: We have already given you what you wanted. You have won, remember? Now please go away. There is nothing more that we can do.

Member/Citizen: But I am hungry!

State/Society: Shut up.

(*Member/Citizen mutely attempts to swallow the judgment in her favor.*)[22]

In a more thoughtful way than 'shut up', a former Black Consciousness movement revolutionary leader, Mamphela Ramphele (a managing director at the World Bank during the early 2000s and later a wealthy venture capitalist), argued forcefully against the rights-based strategy, for it soon becomes a classic culture of entitlement:

The whole approach of the post-apartheid government was to deliver free housing, free this, free the other. This has created expectations on the part of citizens, a passive expectation that government will solve problems. It has led to a 'disengaged citizenry' coupled with a style of leadership in the previous administration that neither accommodated nor welcomed criticism. Thus when people's expectations are not met, they revert to the anti-apartheid mode of protest which is destroy, don't pay, trash. We are yet to grasp the role of citizens as owners of democracy.[23]

The same week, deputy police minister, Fikile Mbalula, alleged:

We have just established recently that in actual fact, there is an element of criminality perpetrated by aboTsotsi [bandits] within our communities who have other intentions not related to service delivery, but use service delivery protests as a tool to commit their intended crime.[24]

Ramphele and Mbalula were amongst many who criticised activists demanding water rights. Yet the activists refused to disengage, and instead continued to protest vigorously, at one of the world's highest per capita rates. Police recorded an average of more than 800 protests annually that they termed 'dissatisfaction with service delivery' from 2009 to 2012.[25] That rate reflected a steady 40 to 45 per cent dissatisfaction level identified in polls undertaken by the Human Sciences Research Council (HSRC) from 2003

22   Ibid 816–7.
23   Cited in P Green '100 Days, 11 Issues' *Mail & Guardian* (17 August 2009).
24   F Mbalula 'Speech Delivered at the Nelson Mandela Bay Crime Prevention Summit by the Deputy Minister: Police' (13 August 2009).
25   SA Police Services 'Service Delivery Protests: January 2009 – November 2012' <http://www.scribd.com/doc/121609151/Service-Delivery-Protests-Official-SA-stats>.

to 2011.[26] Moreover, the strategy of refusing to pay for water and electricity proved to be effective in pushing the state to make concessions such as the 2000 ANC Free Basic Water promise and the 2008 expansion of the free water allocation in Johannesburg, Durban and a few other cities.

But the state's overall objective has been to define rights-based protest (and indeed all service delivery protest, along with individual solutions such as informal water or electricity reconnection following disconnections) as illegitimate, and instead to channel the radical language of grassroots activists towards the courts. According to Danie Brand, 'The law, including adjudication, works in a variety of ways to destroy the societal structures necessary for politics, to close down space for political contestation'. Brand specifically accuses courts of 'domesticating issues of poverty and need' so that they become depoliticised, 'cast as private or familial issues rather than public or political'.[27]

V   FROM RIGHTS TO COMMONS, FROM WATER TO CLIMATE

Karen Bakker notes a variety of problems associated with a narrative of human rights applied to water:

> The adoption of human rights discourse by private companies indicates its limitations as an anti-privatization strategy. Human rights are individualistic, anthropocentric, state-centric, and compatible with private sector provision of water supply; and as such, a limited strategy for those seeking to refute water privatization. Moreover, 'rights talk' offers us an unimaginative language for thinking about new community economies, not least because pursuit of a campaign to establish water as a human right risks reinforcing the public/private binary upon which this confrontation is predicated, occluding possibilities for collective action beyond corporatist models of service provision.[28]

Based on the experiences in the Johannesburg water conflicts, the most logical route through and beyond the limitations intrinsically imposed by rights-based strategies is *not* to defend the Right to Water at Rio+20. The alternative is to explore and shift advocacy efforts towards a 'commons' strategy and indeed an entire culture of sharing, of 'ubuntu' that cuts against the grain of individualised liberties and their potential cooptation within a Green Economy regime. According to the 'onthecommons' website:

> The commons is a new way to express a very old idea-that some forms of wealth belong to all of us, and that these community resources must be actively protected and managed for the good and all. The commons are the things that we inherit and create jointly, and that will (hopefully) last for generations to come. The commons consists of gifts of nature such as air, oceans and wildlife as well as shared social creations such as libraries, public spaces, scientific research and creative works.[29]

---

26   HSRC 'Are you being Served? Perceptions of Service Delivery' *HSRC Review* (2012) <http://www.hsrc.ac.za/HSRC_Review_Article-325.phtml>.

27   D Brand 'The Politics of Need Interpretation and the Adjudication of Socio-Economic Rights Claims in South Africa' in AJ van der Walt (ed) *Theories of Social and Economic Justice* (2005) 17–35.

28   K Bakker 'The "Commons" versus the "Commodity"': Alter-Globalization, Anti-Privatization and the Human Right to Water in the Global South' (2007) 39 *Antipode* 430, 447.

29   See <http://onthecommons.org/content.php?id=1467>.

For Michael Hardt:

> On the one hand, the common refers to the earth and all of its ecosystems, including the atmosphere, the oceans and rivers, and the forests, as well as all the forms of life that interact with them. The common, on the other hand, also refers to the products of human labor and creativity that we share, such as ideas, knowledges, images, codes, affects, social relationships, and the like.[30]

The difference in the two discourses is not merely that water is demanded as an individualised consumption norm in one (rights) and is 'shared' in the other (commons). Other contrasts between the political cultures of rights and of commons are explicitly analysed by Bakker, who insists rights advocates suffer a 'widespread failure to adequately distinguish between different elements of neoliberal reform processes, an analytical sloppiness that diminishes our ability to correctly characterize the aims and trajectories of neoliberal projects of resource management reform'.[31] The rebuttal from Johannesburg activists is that rights discourses – even as purely rhetorical demands for a constitutional entitlement, used to empower ordinary people – can serve as a step towards the commons narrative.

This debate has recurred over centuries of social resistance to commodification and 'enclosure'). Today, Bakker suggests, the water sector includes 'alterglobalization' movements engaged in the construction of alternative community economies and cultures of water, centred on concepts such as the commons and 'water democracies'. A crucial missing element in the rights discourses is environmental, Bakker insists: 'The biophysical properties of resources, together with local governance frameworks, strongly influence the types of neoliberal reforms which are likely to be introduced'. Bakker is concerned that 'in failing to exercise sufficient analytical precision in analyzing processes of "neo-liberalising nature", we are likely to misinterpret the reasons for, and incorrectly characterize the pathway of specific neoliberal reforms'.[32]

This insight, in turn, should generate concern about climate activist narratives, in part because within the Climate Justice movement there is a tendency to draw upon rights-related narratives, especially for generational rights, or 'Greenhouse Development Rights'.[33] Where did Climate Justice begin, and do the dangers of rights talk apply? A conference in The Hague sponsored by the New York group CorpWatch in 2000 was the first known event based on the term Climate Justice.[34] Four years later, the Durban Group for Climate Justice was launched, and for many years remained an important strategic listserve for those opposed to the neo-liberal strategy of carbon trading.[35] The sometimes inchoate advocacy movement known as Climate Justice Now! (CJN!) began in 2007, and played a role in grassroots environmental advocacy

---

30 M Hardt 'Politics of the Common' Contribution to the Reimagining Society Project hosted by ZCommunications (6 July 2009).
31 Bakker (note 28 above) 436.
32 Ibid.
33 P Bond *Politics of Climate Justice* (2012).
34 See J Karliner 'Climate Justice Summit Provides Alternative Vision' CorpWatch <http://www.corpwatch.org/article.php?id=977; and see http://www.corpwatch.org/article.php?id=1048> for the first definition I have seen, dating to late 1999.
35 See <http://www.durbanclimatejustice.org/>.

as well as global-scale UN climate summits.[36] The highest-profile of these, with 100,000 protesters demanding a strong agreement from negotiators, was in Copenhagen in 2009; contesting mainstream environmentalists, Danes and other Europeans formed a Climate Justice Alliance whose 'Reclaiming Power' protest was severely repressed by Copenhagen police.[37]

Shortly after the Copenhagen summit's well-recognised failure, the Bolivian government led by Evo Morales and his then UN ambassador, Pablo Solon, hosted a 2010 conference in Cochabamba, attended by 35,000 activists, including 10,000 from outside the country.[38] This was important partly because of attempts to incorporate within Climate Justice politics a commitment to carbon markets and offset payments, especially through the Reducing Emissions from Deforestation and Forest Degradation (REDD) programme. The Cochabamba conference adopted several demands that were anathema to mainstream climate politics, and that borrowed the language of rights:

- 50 per cent reduction of greenhouse gas emissions by 2017;
- acknowledging the climate debt owed by developed countries;
- full respect for human rights and the inherent rights of indigenous people;
- universal declaration of rights of Mother Earth to ensure harmony with nature;
- establishment of an International Court of Climate Justice;
- rejection of carbon markets and commodification of nature and forests through the REDD programme;
- promotion of measures that change the consumption patterns of developed countries;
- end of intellectual property rights for technologies useful for mitigating climate change; and
- payment of six per cent of developed countries' GDP to addressing climate change.[39]

REDD proved amongst the most important wedge issues within the Climate Justice community, for late in 2010, sharp controversies emerged over forest preservation as major US environmental foundations attempted to resurrect market strategies.[40] Nevertheless, from the realisation that 'neo-liberalised nature' was the new global-governance approach for environmental (and social) management, there emerged, in direct response, a new Climate Justice philosophy and ideology, principles, strategies and tactics. The question still to be answered is whether transcending rights talk in favour of commoning is the appropriate route forward for climate activism; at least with respect

---

36   See <http://www.climate-justice-now.org/>; and W Kaara 'Reclaiming Peoples' Power in Copenhagen 2009: a Victory for Ecosocialist Ecofeminism' (2010) 21 *Capitalism Nature Socialism* 107.

37   See <http://www.climate-justice-action.org/>.

38   See <http://pwccc.wordpress.com/>.

39   See <http://cochabamba2010.typepad.com/blog/2010/08/the-proposals-of-peoples-agreement-in-the-texts-for-united-nations-negotiation-on-climate-change.html>.

40   Movement Generation et al 'Open Letter to 1 Sky' Oakland (2010).

to linkage of issues through a commons perspective, there are enormous potentials, as Edgardo Lander explained in his review of the Cochabamba:

> Struggles for environmental or climate justice have managed to bring together most of the most important issues/struggles of the last decades (justice/equality, war/militarization, free trade, food sovereignty, agribusiness, peasants' rights, struggles against patriarchy, defense of indigenous peoples' rights, migration, the critique of the dominant Eurocentric/colonial patterns of knowledge, as well as struggles for democracy, etc, etc). All these issues were debated on Cochabamba and, to some degree, present in the Cochabamba Peoples' Agreement.[41]

VI CONCLUSION

Does the eco-social critique of the limited effect of rights apply to South African water-rights activists and does it explain the constraints associated with their human rights discourse? At the same time, does it offer lessons for Climate Justice activism, as to avoiding similar limitations? Perhaps most importantly, in order to make their case for more water without prepayment meters, the Soweto activists and their lawyers focused only upon the consumption needs of low-income residents. Hence several other processes were downplayed: the source of a large amount of Johannesburg's water in the Lesotho dams; the manner in which Rand Water – the catchment management agency between the dams and Johannesburg – processed and distributed the water; the financing of the bulk system through the World Bank and other creditors; the extremely high consumption norms of Johannesburg's wealthier residents and large corporations; and the disposal of water through the system's sanitation grid into a water table and groundwater beset by ecological crises. In other words, linkage of issues was lost by virtue of the narrow human-centric channelling that legalistic rights talk compels of water activists. This is not to say that by adding natural rights, the human rights case would be strengthened, of course. The potential to invoke a 'water reserve' (ie letting a river flow all the way to the ocean) within national legislation (for example the National Water Act) does not necessarily assist poor people in gaining access to water, without first challenging the extreme abuse of water by corporations (Eskom is the most wasteful given its use of cooling water for coal-fired power plants), golf courses, timber plantations and wealthy households with swimming pools and English gardens.

Adding environmental factors is only the first step to 'commoning' water. Much more important is establishing a base amongst water consumers for a different way of arranging water distribution and disposal. For if done without adequate foresight, Bakker warns:

> appeals to the commons run the risk of romanticizing community control. Much activism in favour of collective, community-based forms of water supply management tends to romanticize communities as coherent, relatively equitable social structures, despite the fact that inequitable power relations and resource allocation exist within communities.[42]

---

41   E Lander 'Reflections on the Cochabamba Climate Summit' <http://www.tni.org/article/reflections-cochabamba-climate-summit>.
42   Bakker (note 28 above).

The challenge, thus, is to introduce a strong culture of water commons as an ideology, so that public consciousness and daily life are suffused with the vision of equitable access ensured through collective action in a context of ecological limits. That will serve as an antidote to the 'neo-liberal populism' that may well emerge to re-commodify commons processes. For example, faddish techniques of micro-financing and 'self-help' entrepreneurial ideologies drawing on a 'culture of social entrepreneurship' are now applied to public goods such as water and health care.[43] In the name of the 'right to credit' based on breathing the life of finance into 'dead capital', there has been enormous damage done to a commons of social trust. This is true even in the case of Muhammad Yunus' Grameen Bank, given that micro-credit is now increasingly held responsible for thousands of small-farmer suicides in South Asia and other manifestations of market/society failure. One of the most influential micro-entrepreneur advocates, Hernando de Soto, rests his vision of property rights upon the collateralisation of land, shacks, livestock and other goods informally owned by poor people (but currently dead capital), all the better to invoke micro-finance and in turn an often mythical successful rise to market-based wealth generation. Such capture of commons processes at local level should be contrasted with the changes required at the national scale, and potentially globally once the balance of power improves, to fundamentally redirect our inherited patterns of extraction, production, distribution, transport, financing, consumption and disposal.

Even greater challenges can be found along similar lines when it comes to climate change, a problem that amplifies the need for radical change in all the inherited systems that have proven so destructive. This is true especially as absolute water scarcity emerges, for countries like South Africa – and metropolitan areas such as Johannesburg – will become sites of conflict thanks to climate change, paralleling rural Darfur, Sudan, where sustained drought catalysed a ferocious war over land and water access. This kind of disaster brings us, finally, to recall that in 2009 in Addis Ababa, Ethiopia, the African Union demanded wealthy industrialised countries pay reparations for damage done by climate change under the rubric of 'climate debt'. Numerous other forms of ecological debt could be calculated and paid for by over-consumers in the Global North, given that the perspective required to move in this direction is to understand the web of life connecting North and South as a great commons.[44]

In the water sector, activist awareness of the ecological aspects of water as commons is growing especially because of climate change. The Johannesburg region is crucial because it is the most intensive site for (non-smelting) electricity usage in South Africa, its water tables are being ruined through Acid Mine Drainage, its main resource (gold) is nearly exhausted,

43   P Bond 'Microcredit Evangelism, Health and Social Policy' (2007) 37*International J of Health Services* 229.
44   'Rich Countries Owe Poor a Huge Environmental Debt' *Guardian* (18 January 2008); also see K Sharife & P Bond 'Payment for ecosystem services versus ecological reparations: The "green economy"', litigation and a redistributive eco-debt grant' (2013) in this volume.

and its manufacturing base is uncompetitive with imports from East Asia. As a financial and services centre it has thrived, but the sustainability of such activity is limited given the country's vast problems with current account balances, foreign debt and an unstable currency. Moreover, it is a city with vast reservoirs of conscientised activists in civil society, whose honeymoon with the South African state after apartheid was very short indeed.

What was and is necessary, for exploration in Johannesburg after Phiri, or in Durban after COP17, or anywhere influenced by the 2012 Rio+20 Summit, as well as in all the future sites of struggle over water and environmental services across the world, are new ideas and strategies that can transcend consumption-based rights demands. As a first step, we need more coherent critiques of the full range of practices that undermine our ability to perceive and respect water and other aspects of nature as a commons. These strategies may emerge through fusions of community, environmental and labour in the alliance-formation that necessarily occurs during eco-social justice struggles, as rights-talk meets its limits, and as the commons appears as a new frontier.

PAYMENT FOR ECOSYSTEM SERVICES VERSUS ECOLOGICAL REPARATIONS: THE 'GREEN ECONOMY', LITIGATION AND A REDISTRIBUTIVE ECO-DEBT GRANT

KHADIJA SHARIFE AND PATRICK BOND

INTRODUCTION

According to Achim Steiner, director of the United Nations (UN) Environment Programme, until the value of what he terms 'ecosystem services' are acknowledged, ecological free-riding will never be slowed: 'An intact hectare of mangroves in a country like Thailand is worth more than US$1,000. Converted into intensive farming, the value drops to an estimated US$200 a hectare and the same for aquaculture.'[1] By extending the logic of capitalism from money and labour to oceans and forests, promoters of the 'Green Economy' such as the World Wide Fund for Nature (WWF), the Organisation for Economic Cooperation and Development (OECD), Conservation International and the World Bank hope that nature can finally 'invoice' natural capital users. As WWF president Yolanda Kakakabadse argued, 'Until now, natural wealth or capital has been considered as global commons and therefore treated as a free good ... WWF supports multilateral, corporate and academic entities as a force behind the valuation of natural capital ...'.[2] Added Serge Tomasi, deputy director of the OECD's Development Co-operation Directorate, 'Carbon markets and other market instruments could help to fix the right price to the natural resources, and to integrate environmental degradation externalities into decision making processes'.[3] In May 2012, ten African governments endorsed the Gaborone Declaration, where proper accounting of natural capital was declared to underpin sustainability, growth and poverty reduction.[4]

But there are crucial questions: will this framing form the basis for effective campaigning to 'make the polluter pay' from the standpoint of juridical accountability? Who does the invoicing on behalf of nature? Should nature be invoiced at all, or is pricing inherently misguided, eliding intrinsic value? Who defines 'intrinsic'? Are there specific instances in which nature values

---

1   K Sharife 'What is the Real Value of Africa's Wealth?' *African Business* (May 2012).
2   Interview with K Sharife, Rio de Janeiro, Brazil *The Africa Report* (June 2012).
3   Ibid.
4   Conservation International 'Gaborone Declaration Pioneers Commitment to Value Natural Capital' (May 2012) <http://www.conservation.org/newsroom/pressreleases/Pages/Gaborone-Declaration-Pioneers-Commitment-to-Value-Natural-Capital-.aspx>.

can be assessed as prices in order to achieve restorative environmental justice? If monetized, can conversions from values to prices be determined by those on the frontline of climate change, whose very right to life is threatened, and who are owed an 'ecological debt' to compensate for the destruction – as of December 2012 termed 'loss and damage' in the UN climate negotiations – that they have suffered? Or will the design come from those seeking 'payment for ecosystem services' such that nature becomes fully commodified and subject to financial market whims? To what extent would the inclusion of those suffering loss and damage be located in a nominal manner, empowered only enough to legitimate capitalist calculative entities? Should ecological debt be monetized as one complementary vehicle towards reparation, or does it fall squarely within the financialisation trap to 'make markets'? How are markets in nature – such as carbon trading – working at present? And if there are consistent failures, what practical alternatives are there for halting ecological destruction (especially climate change) and compensating victims, outside the realm of the market?

The answers to these questions are becoming urgent, for the stakes could not be higher. In 2006, Christian Aid estimated that 182-million Africans were at risk of premature death due to climate change this century.[5] Seven years later, much more dire predictions about climate change are usually offered by scientists. The way that both mitigation and adaptation narratives are unfolding poses a great threat to the victims, since the valuation of life and planetary ecology through market mechanisms will punish those without market access many times over. This is especially true in Africa where resource-curse mechanisms leave those with political power closely overlapped with those whose extractive economic interests are opposed to environmental justice. That means that even when court-based justice is sought, in the form of an ecological debt payment, the danger emerges that political elites and other rentiers will capture the adaptation funding. The alternative is that the funds go directly to victims of ecological damage, in the form of a Basic Income Grant (BIG) piloted in Namibia. Before making this case, we first consider why current trends in *pricing* of nature have set the stage for systemic denial of eco-social rights.

II   VALUES VERSUS PRICES AT THE RIO+20 EARTH SUMMIT

The idea of 'pricing' human and ecological life has been present in one form or another ever since slavery, and the origins of systems such as insurance and tort law. The value of one European was the equivalent of ten Chinese in a 1994 Intergovernmental Panel on Climate Change (Working Group 3) report, which led India's minister for Environment, Kamal Nath, to vehemently reject the 'absurd and discriminatory global cost-benefit analysis procedures

---

5   Christian Aid 'Developing Countries Demand Compensation for Climate Change' (2008) <http://
     www.christianaid.org.uk/pressoffice/pressreleases/august2008/climate_change_talks_accra.
     aspx>.

propounded by the economists'.[6] Valorising people and the planet became more important for capital after the world economic crisis worsened after 2008. The turn to 'Green Economy' rhetoric looms as 'accumulation through dispossession', in the words of David Harvey.[7] This attempt to rationalise environmental management (also known as 'ecological modernisation') represents a potential saviour for footloose financial capital, and is particularly welcome to those corporations panicking at market chaos in the topsy-turvy fossil-fuel, water, infrastructure construction, technology and agriculture sectors.

As we see in more detail below, the Rio+20 Earth Summit of June 2012 provided a renewed official faith in market mechanisms, following the logic of two South African precedents: the 2002 World Summit on Sustainable Development in Johannesburg (Rio+10) and the December 2011 Durban Conference of the Parties 17 (COP17) climate summit.[8] At all three sites, the chance to begin urgent environmental planning to reverse ecosystem destruction was ignored and instead, big- and medium-governments' negotiators acted on behalf of their countries' corporations, with resulting increased pollution and privatisation of nature.

To illustrate, Durban's main winners appeared to be those from Washington who had come intent on halting progress. 'The Durban Platform was promising because of what it did not say,' remarked Trevor Houser, a former top aide to chief US State Department climate negotiator Todd Stern. Speaking to *The New York Times* a few weeks later at the Davos World Economic Forum in Switzerland, Houser bragged, 'There is no mention of historic responsibility or per capita emissions. There is no mention of economic development as the priority for developing countries. There is no mention of a difference between developed and developing country action'.[9]

The attitude of Washington powerbrokers really has not changed in 20 years, as we can ascertain by tracing back to the most infamous statement of US self-interest in global ecological governance, by Larry Summers. He was, at the time, World Bank chief economist, but soon would become a top-ranking Clinton administration official, rising to finance minister status until the 2000 'election' in Florida ended Democratic Party control of the White House. In December 1991, as the World Bank prepared to take over financing major functions related to the Rio Earth Summit, chief economist Summers signed a bizarre memo to his closest Bank colleagues suggesting, in effect, that nature be privatised, to better assess costs and benefits of Bank ecological intervention. As he put it, 'I've always thought that under-populated countries

---

6   C Okereke *Global Justice and Neoliberal Environmental Justice: Ethics, Sustainable Development and International Co-Operation* (2008) 108.

7   D Harvey 'The "New" Imperialism: Accumulation by Dispossession' in L Panitch & C Leys (eds) *Socialist Register 2009* (2009) <http://socialistregister.com/index.php/srv/article/view/5811>.

8   See P Bond *Unsustainable South Africa: Environment, Development and Social Protest* (2002); P Bond *Politics of Climate Justice* (2012).

9   M Levi 'Another Perspective on the Durban Climate Talks' (2011) <http://blogs.cfr.org/levi/2011/12/16/another-perspective-on-the-durban-climate-talks/>.

in Africa are vastly UNDER-polluted'.[10] Though extremist, such ideology was endorsed by *The Economist* magazine, which leaked the memo in early 1992. The underlying philosophy informed a great deal of Bank and even UN policy ever since. The bottom line was US president George Bush Sr's pronouncement at the Rio Earth Summit: 'The American way of life is not negotiable.'[11] This paved the way for Rio+10 in Johannesburg. At that 2002 World Summit on Sustainable Development, ever more aspects of nature would be seen as 'economic goods'. For example, water commodification by then was the subject of intense conflict, especially over municipal commercialisation. Soweto was one of the world's most publicised water wars, with the Anti-Privatisation Forum's community activists regularly destroying pre-payment meters and demanding a doubling of the Free Basic Water supply. In Johannesburg, the huge Paris water company Suez found the going tough and instead of managing outsourced municipal services for an anticipated 30 years, left after just five, in 2006. Simultaneously, Suez was in deep trouble across the Third World, losing all its Argentine revenue when activists pressured its leaders to default on profit repatriation agreements in 2002.[12]

Notwithstanding such concrete difficulties in 'neo-liberalising nature', as this process is increasingly termed, global climate policy debates have not shifted much since 1997, when US vice-president Al Gore went to Japan's COP3 in Kyoto, promising that Washington would sign the climate Protocol if it included carbon markets as an escape hatch for companies that polluted too much and then wanted the right to buy other companies' pollution permits. The markets were granted, but the US senate voted 95-0 against endorsing Kyoto. As we argue below, the results of the emissions trading experiment have been extremely disappointing.

In preparation for Rio+20, the April 2012 World Bank report, *Inclusive Green Growth,* argues, 'Care must be taken to ensure that cities and roads, factories, and farms are designed, managed, and regulated as efficiently as possible to wisely use natural resources while supporting the robust growth developing countries still need'.[13] Bank staff led by Inger Andersen and Rachel Kyte aim to move the economy 'away from suboptimalities and increase efficiency – and hence contribute to short-term growth – while protecting the environment'. In this narrative, certain uses of resources are off limits for polite discussion, as Bank staff dare not question financiers' commodity speculation, export-led growth or the irrationality of so much international trade, including wasted bunker fuel for shipping, not to mention truck freight. Yet the Bank cannot help but momentarily inject a power variable into its technicist analysis:

> That so much pricing is currently inefficient suggests complex political economy considerations. Whether it takes the form of preferential access to land and credit or access to cheap energy and

10   L Summers 'The Bank Memo' (1991) <http://www.whirledbank.org/ourwords/summers.html>.
11   J Vidal 'Rio+20: Earth Summit Dawns with Stormier Clouds than in 1992' (19 June 2012) <http://www.guardian.co.uk/environment/2012/jun/19/rio-20-earth-summit-1992-2012>.
12   P Bond *Talk Left, Walk Right* (2006).
13   World Bank *Inclusive Green Growth* (2012) 49.

resources, every subsidy creates its own lobby. Large enterprises (both state owned and private) have political power and lobbying capacity. Energy-intensive export industries, for example, will lobby for subsidies to maintain their competitiveness.[14]

Would the Bank practice what it preaches about ending 'inefficient' subsidisation, given how it amplifies irrational power relations when maintaining the world's largest fossil-fuel financing portfolio? When *Inclusive Green Growth* argues that 'Governments need to focus on the wider social benefits of reforms and need to be willing to stand up to lobby groups',[15] South Africans cannot forget the Bank's own largest-ever project credit, granted in April 2010. The US$3.75-billion loan for a 4800 MW coal-fired power plant at Medupi was, according to former Bank president Robert Zoellick and his colleagues, aimed at helping poor South Africans. In reality the benefits are overwhelmingly to mining houses, which get the world's cheapest electricity (less than US$0.02 per kiloWatt hour). The costs of Medupi and its successor Kusile are borne not just by all who will suffer from climate change. All South Africans are losing access to electricity through disconnections, and as a result, engaging in world-leading rates of community protest because to pay for Medupi and Kusile, price increases exceeded 130 per cent between 2008 and 2012 (to US$0.15 per kiloWatt hour).[16]

The Bank's *Inclusive Green Growth* arguments always return to profit incentives: 'If the environment is considered as productive capital, it makes sense to invest in it, and environmental policies can be considered as investment.'[17] Facing up to pollution externalities is deceptively simple within the Bank's pre-existing neoliberal narrative, of fixing a market problem with a market solution. For example, 'Lack of property rights in the sea has led to overfishing – in some cases with devastating results. The use of individual transferable quotas can correct this market failure, increasing both output and employment in the fishing industry'.[18] The Bank's reversion to transferable 'cap-and-trade' quotas is most extreme in the greenhouse gas (GHG) markets, where its writers fail to acknowledge profound flaws that have crashed the price of a tonne of carbon from €35 to €7 these last six years. The Bank, which subsidises carbon trading, mentions only a few allegedly-fixable European Union (EU) Emissions Trading Scheme design problems, but ignores the deeper critique of carbon markets.

Likewise, the UN Environment Programme came to view 'the sustainability crisis as the biggest-ever "market failure"'. The directors of the Barcelona-based Environmental Justice Organisations Liabilities and Trade project, Joan Martinez-Alier and Joachim Spangenberg, issued a statement about the Green Economy at Rio, 'Describing it this way reveals a specific kind of thinking: a market failure means that the market failed to deliver what in principle it could have delivered, and once the bug is fixed the market will solve the problem'.

14   Ibid.
15   Ibid.
16   P Bond (ed) *Durban's Climate Gamble* (2011).
17   S Hallegatte, G Heal, M Fay & D Tréguer 'From Growth to Green growth' VOX (24 March 2012) <http://www.voxeu.org/article/growth-green-growth>.
18   *Inclusive Green Growth* (note 13 above) 38.

They reverse this logic: '[U]nsustainable development is not a *market failure* to be fixed but a *market system failure*: expecting results from the market that it cannot deliver, like long-term thinking, environmental consciousness and social responsibility.'[19]

Under the rubric of the Green Economy, corporations are seeking new technological 'False Solutions' to the climate and other environmental crises, including dirty forms of 'clean energy' (nuclear, so-called 'clean coal', fracking 'natural gas', hydropower, hydrogen, biofuels, biomass and biochar, carbon capture and storage experiments) and other geoengineering strategies such as Genetically Modified trees as plantations to sequester carbon, sulfates in the air to shut out the sun, iron filings in the sea to create algae blooms, and large-scale solar reflection such as industrial-scale plastic-wrap for deserts.

III FROM AFRICAN 'NATURAL CAPITAL' TO PRICING TO MARKETS

These dubious tactics aside, the philosophical underpinning of the Green Economy needs wider questioning. The precise wording is terribly important, as Africans began to understand after the Gaborone Declaration. In May 2012, Botswana's president Ian Khama brought together leaders from nine other African countries – Gabon, Ghana, Kenya, Liberia, Mozambique, Namibia, Rwanda, South Africa and Tanzania – to 'quantify and integrate into development and business practice' what ordinary people consider to be the innate *value* of nature. But these leaders and their conference sponsor Conservation International mean something else, devoid of eco-systemic, spiritual, aesthetic, and intrinsic qualities. The Declaration insists:

> Watersheds, forests, fisheries, coral reefs, soils, and all natural resources, ecosystems and biodiversity constitute our vital natural capital and are central to long-term human well-being, and therefore must be protected from overuse and degradation and, where necessary, must be restored and enhanced.[20]

By relegating the environment to mere natural capital, the next step is to convert value into *price* and then sell nature on the market. All manner of financialisation strategies have emerged to securitise ecosystem services, most obviously in carbon markets which continue failing miserably to deliver investor funds to slow climate change, as discussed below. But environmentally-oriented bankers are not deterred. Explained City of London investor Simon Greenspan, whose firm Tullett Brown won *World Finance* magazine's 'Western European Commodities Broker of the Year' award in March 2012:

> At Tullett Brown we've only ever invested in areas of the market that have truly stood the test of time, such as gold and silver and property. When our analysts were looking for the next great area of growth it was fairly obvious to them. It was the planet, it was the environment.[21]

---

19 Environmental Justice Organisations Liabilities and Trade 'No Green Economy without Environmental Justice!' (June 2012) <http://www.ejolt.org/2012/06/no-green-economy-without-environmental-justice/>.
20 Conservation International (note 4 above).
21 C Lang 'Carvier Limited: "3 Million Units Available!!" from Brazil – the Ethical Alternative to Carbon Credits?' (6 June 2012) <http://www.redd-monitor.org/2012/06/06/carvier-limited-3-million-units-available-from-brazil-the-ethical-alternative-to-carbon-credits/>.

(Just days later, British financial authorities forced Tullett Brown into provisional liquidation followed by litigation regarding commercial fraud.)

Reacting to the Gaborone Declaration, Nnimmo Bassey from the Niger Delta non-governmental organisation (NGO) Environmental Rights Action and Friends of the Earth International warned, 'The bait of revenue from natural capital is simply a cover for continued rape of African natural resources'. Thanks to inadequate protection against market abuse, he adds, 'The declaration will help corporate interests in Rio while impoverishing already disadvantaged populations, exacerbate land grabs and displace the poor from their territories'.[22] And to further illustrate the pernicious way markets undermine nature, Zimbabwe's president Robert Mugabe would say of the rhino and elephant in 1997, 'The species must pay to stay' – which in turn allowed him and wealthy (white) game farm owners to offer rich overseas hunters the opportunity to shoot big game at high prices.[23] The dilemma about hunt marketing is that it doesn't stop there: black markets in rhino horns and elephant tusks are the incentive for poachers to invade not just poorly defended game parks north of the Limpopo River, but also now in South Africa where in 2013 the prospect of formal markets in rhino horns and elephant tusks re-emerged. The alternative strategy would have been to tighten the Convention on International Trade in Endangered Species' (CITES) restrictions against trade in ivory. But because South Africa's game-farm owners and free-market proponents influenced Pretoria to press for relaxation of CITES' ban, hundreds of elephant and rhino corpses denuded of horns and tusks now litter the bush each year.

At best, the Gaborone Declaration commits the ten countries to 'reducing poverty by transitioning agriculture, extractive industries, fisheries and other natural capital uses to practices that promote sustainable employment, food security, sustainable energy and the protection of natural capital through protected areas and other mechanisms'.[24] How, though, is the crucial question, for the system relies upon intrinsic nature as applicable to saving, sustaining and protection of the environment but only when converted into capital.

Boiling down a complex argument about how to properly value people and nature from her book *Eco-Sufficiency & Global Justice*, University of Sydney-based political ecologist Ariel Salleh observes how a triple externalisation of costs 'takes the form of an extraction of surpluses, both economic and thermodynamic: (1) a social debt to inadequately paid workers; (2) an embodied debt to women family caregivers; and (3) an ecological debt drawn on nature at large'.[25] At minimum, addressing these problems requires full-fledged re-accounting to toss out the fatally-flawed gross domestic product

22   P Bond 'At Rio+20: Values versus Prices' (2012) <http://roadlogs.rio20.net/at-rio20-values-versus-prices/>.

23   F Macleod 'Africa must Pay for its Wildlife' *Mail & Guardian* (26 September 1997) <http://mg.co.za/article/1997-09-26-africa-must-pay-for-its-wildlife>.

24   Conservation International (note 4 above).

25   A Salleh 'Women, Food Sovereignty and Green Jobs in China' (2012) <http://www.foe.org.au/women-food-sovereignty-and-green-jobs-china>.

(GDP) indicator, and to internalise environment and society in the ways we assess costs and benefits.

Of course, this exercise would logically both precede and catalyse a full-fledged transformation of financing, extraction, production, transport and distribution, consumption and disposal systems. Indeed, if one makes just four simple corrections to GDP, as does the World Bank in *The Changing Wealth of Nations,* there suddenly emerges an unintended consequence of neo-liberalised nature. While three of these factors are relatively minor – depreciation of fixed capital, education spending and pollution – the fourth is foundational in its ideological implications: calculating the lost 'natural capital' in the form of non-renewable resource depletion associated with extractive industries.

This indeed is the critical point. When non-renewable resources are dug out of the soil, there should logically be a permanent *debit* against genuine national savings (due to the consequent decline in a country's natural resources); instead national income accounting provides a *credit* to GDP. Thus in many situations, once the calculations are made, *it becomes logical to leave resources in the ground.* This is especially true in sub-Saharan Africa, because since the commodity boom began in the early 2000s, according to *The Changing Wealth of Nations,* Africa has suffered negative genuine savings mainly because of non-renewable resource decay in the context of resource-cursed neo-colonial politics. Such resource decay would be calculated at an even greater rate if such reports would take into account the illicit flight of resource revenue. Even without that critical additional factor this stripping of African wealth – seen via 'adjusted net savings' year on year – had reached six per cent of gross national income by 2008, according to the Bank.[26]

But there is a need for measurement and valuation. In their seminal *Ecological Economics* article, Mathis Wackernagel et al write:

> To translate the strong sustainability criterion into concrete numbers and to examine whether society lives within its ecological capacity, a first overview needs to account for natural capital and its uses at the national and global level ... Ecological footprint calculations are based on two simple facts: first, we can keep track of most of the resources we consume and many of the wastes we generate; and second, most of these resource and waste flows can be converted to a biologically productive area necessary to provide these functions. Thus, ecological footprints show us how much nature nations use.[27]

The report widely credited with putting 'natural capital' on the negotiating table is the UN Environment Programme study, *The Economics of Ecosystems and Biodiversity* (TEEB), led by former Deutsche Bank official Pavan Sukhdev:

> Nature comprises around 45-90 per cent of the GDP of the poor in some developing countries. By assigning economic values to the services flowing from nature to people, policy makers and the global economy can start to account for the costs of biodiversity loss, as well as reward responsible custodians for the benefits that natural ecosystems provide.[28]

26   World Bank *The Changing Wealth of Nations* (2011).
27   M Wackernagel, L Onisto, P Bello, A Linares, I Falfan, J Garcia, A Guerrero & G Guerrero 'National Natural Capital Accounting with the Ecological Footprint Concept' (1999) 29 *Ecological Economics* 375 <http://www.sciencedirect.com/science/article/pii/S0921800998900635>.
28   K Sharife 'Who Should Pay for Climate Damage?' *African Business* (March 2013).

Sukhdev acknowledged that valuation was a human system and that intrinsic value would not necessarily be included.

IV   CARBON CAPITAL

It should now be morally obvious that a vast debt is owed Africa, in view of this process of what might be considered 'looting'.[29] Moreover, Africa's odious debt has been structurally tied to the looting of resources. But instead of being properly compensated for social, economic and environmental damage, Africa is being drawn into an environmental policy framework and climate finance regime based in part upon failing financial markets that have mainly enriched speculators and impoverished the continent's poor people. In the wake of South Africa's unsuccessful hosting of the December 2011 Durban COP17 climate summit, where negotiators again postponed decisions to save the planet from catastrophic warming and ever more extreme weather events, the newest signals from the UN, World Bank and EU suggest that rising fears of carbon markets in Africa are well grounded.

Instead, those who followed the Durban UN Framework Convention on Climate Change (FCCC) COP17 in December 2011 heard that the solution to climate crisis must centre on markets, in order to 'price pollution' and simultaneously cut the costs associated with mitigating GHGs. Moreover, say proponents, these markets are vital for funding not only innovative carbon-cutting projects in Africa, but also for supplying a future guaranteed revenue stream to the Green Climate Fund (GCF), whose design team co-chair, Trevor Manuel (South Africa's planning minister), argued as early as November 2010 that up to half of the GCF revenues would logically flow from carbon markets.

If we take this logic seriously, of most interest for Africans is one small but important component of the emissions market, the Clean Development Mechanism (CDM). The CDM's size as a percentage of total carbon trading volume has been around just five per cent, and the vast bulk of financing has gone to just four countries, none in Africa. The strategy was established within the Kyoto Protocol in 1997. It aims to facilitate innovative carbon-mitigation and alternative development projects by drawing in funds from northern GHG emitters in exchange for permitting their continuing pollution. CDMs generate Certified Emissions Reductions (CERs) that act as another asset class to be bought, sold and hedged in the market. The EU's Emissions Trading Scheme (ETS) is the main site of trading, following a failed attempt at a carbon tax in Europe prior to 1997, due to intensive lobbying from resistant companies.

CDMs were   created   to   allow   wealthier   countries   classified   as 'industrialised' – or Annex 1 – to engage in emissions reductions initiatives in poor and middle-income countries, as a way of eliding direct emissions reductions. Put simply: the owner of a major polluting vehicle in Europe can pay an African country to not pollute in some way, so that the owner of the vehicle is allowed to continue emitting. In the process, developing

29   *Changing Wealth of Nations* (note 26 above).

countries are, in theory, benefiting from sustainable energy projects. The use of such 'market solutions' will, supporters argue, lower the business costs of transitioning to a post-carbon world. In a cap and trade system, after a cap is placed on total emissions, the high-polluting corporations and governments can buy ever more costly carbon permits from those polluters who do not need so many, or from those willing to part with the permits for a higher price than the profits they make in high-pollution production, energy-generation, agriculture, consumption, disposal or transport. The intent is to remove the 'real seat' of origin, and thus legitimately negate accountability, through financial transaction based on what could arguably be called (speculative) futures markets.

Instead of providing an appropriate flow of climate finance for such transformation, or for projects related to GHG mitigation, the CDM has benefited large corporations (both South and North) and the governments they influence and often control. Many sites of emissions in Africa – for example, methane from rotting rubbish in landfills, flaring of gas from oil extraction, coal-burning electricity generation, coal-to-liquid and gas-to-liquid petroleum refining, deforestation, decomposed vegetation in tropical dams – require urgent attention, as do the proliferation of 'false solutions' to the climate crisis such as mega-hydro power, tree plantations and biofuels. Across Africa, the CDM subsidises all these dangerous for-profit activities, making them yet more advantageous to multinational corporations, which are mostly based in Europe, the United States (US) or South Africa. In turn, these same corporations – and others just as ecologically irresponsible – can continue to pollute beyond the bounds set by politicians especially in Europe, because the ETS forgives increasing pollution in the North if it is offset by dubious projects in the South. But because communities, workers and local environments have been harmed in the process, various kinds of social resistances have emerged, and in some cases met with repression or cooptation through 'divide-and-rule' strategies, as documented in a 2012 study by researchers at the University of KwaZulu-Natal (UKZN) and Dartmouth College.[30]

Durban's COP17 failed to gain commitments for the vital GHG emissions cuts of 50 per cent by 2020, for ensuring the North's climate debt to the South covers the sorts of damages Kofi Annan specified under a 'polluter pays' logic, and for establishing a transition path to a post-carbon society and economy. Even within the very limited, flawed strategy of carbon markets, there were mixed outcomes from the Durban COP17. In spite of Trevor Manuel's efforts to bring emissions trading into the GCF, where it does not belong, and in spite of the UN CDM executive board's decision to allow 'Carbon Capture and Storage' experiments to qualify for funding, the most profound flaw in the existing market was not addressed. Without an ever-lowering cap on emissions, the incentive to increase prices and raise trading volumes disappears. Worse, in a context of economic stagnation in Annex 1 countries, financial volatility and shrinking demand for emissions reduction credits, the

---

30   UKZN CCS and Dartmouth College Climate Justice Project. 'CDMs Cannot Deliver the Money' (April 2012) <http://cdmscannotdeliver.wordpress.com/>.

world faces increasing sources of carbon credit supply in an already glutted market.

Durban left the world's stuttering carbon markets without a renewed framework for a global emissions trading scheme. Durban turned the Kyoto Protocol – which is now applicable to only 14 per cent of world GHG emissions – into a 'Zombie' (walking-dead) because its heart, soul and brain (binding emissions cuts) all died, as former Bolivian ambassador Pablo Solon put it.[31] All that appears to be moving is the stumbling and indeed crashing commitment to CDMs. These markets can be expected to die completely given the failure at Qatar's COP18 to generate more commitments to legally-binding emissions cuts. And judging by Washington's threat, it won't be until 2020 – the COP26 – when the US will review its own targets: the Copenhagen Accord's meaningless three per cent cuts offered from 1990 to 2020. By then it will be too late, because the Kyoto Protocol's mistaken reliance on financial markets means that the period 1997 to 2011 will be seen as the lost years of inaction and misguided financial quackery – when we urgently need the period going forward from 2012 to be defined as an era in which humanity took charge of its future and ensured planetary survival.

Unlike soft and hard tangible commodities such as corn or gold, the carbon credits exist purely on the basis of 'authorisation' on the part of national governments. If 'deauthorised', the entire credit market – and the justification of hundreds of billions of dollars worth of carbon trades – becomes pure fiction. Chances are that methane – yet another consistently named gas – will also soon become a junk asset. To be sure, the fact that the Kyoto Protocol was nominally extended a few years means that CDMs will continue to be traded, even though from 2007 to 2010 the volume of activity fell by 80 per cent. Jonathan Grant, director of carbon markets and climate policy at PricewaterhouseCoopers stated:

> Thanks to Durban, the CDM will live to see another day, but demand for credits for these projects is lacklustre. Carbon markets are expected to stay in the doldrums, because of oversupply in the (European carbon) market as a result of the recession.[32]

According to Barclays Capital's lead carbon researcher, Trevor Sikorski, there are vast surpluses of credits – at least a billion carbon credits.[33]

That problem will be exacerbated by pressure on the voluntary markets from new Reducing Emissions through Deforestation and Forest Degradation (REDD) offsets as well as by the UN executive board's decision to include Carbon Capture and Storage experimentation in CDMs. Together, these factors have wrecked the European market for CDMs. In the words of emissions trading lawyer Rutger de Witt Wijnen in April 2012, 'We all know there are too many [carbon] allowances around, too many credits, too few emissions, too few market players who are willing to make the market'. And worse, he

31   P Solon 'Wolpe Lecture at the University of KwaZulu-Natal, Durban' (2 December 2011) <http://ccs.ukzn.ac.za>.

32   Reuters 'Carbon Markets Still on Life Support after Climate Deal' (12 December 2011) <http://www.reuters.com/article/2011/12/12/us-climate-carbon-idUSTRE7BB0QT20111212>.

33   Ibid.

continued, 'People and companies are leaving the markets; companies are closing their carbon trading desks, the same for law firms and advisors'.[34]

Concerns about the carbon markets have been expressed for many years by civil society opponents of carbon trading. Frustration with CDMs in Africa reached a critical mass as early as 2004 when the Durban Group for Climate Justice gathered for an historic meeting. A global civil society network, the Durban Group[35] was formed to oppose carbon trading in 2004. The critique can be summed up in eight points: (1) the idea of inventing a property right to pollute is effectively the 'privatisation of the air', a moral problem given vast, growing differentials in wealth inequalities; (2) GHGs are complex and their rising production creates a non-linear impact which cannot be reduced to a commodity exchange relationship (a tonne of CO_2 produced in one place is accommodated by reducing a tonne in another, as is the premise of the emissions trade); (3) the corporations most guilty of pollution and the World Bank – which is most responsible for fossil-fuel financing – are the driving forces behind the market, and can be expected to engage in systemic corruption to attract money into the market even if this prevents genuine emissions reductions; (4) many of the offsetting projects – such as mono-cultural timber plantations, forest 'protection' and landfill methane-electricity projects – have devastating impacts on local communities and ecologies, and have been hotly contested in part because the carbon sequestered is far more temporary (since trees die) than the carbon emitted; (5) the price of carbon determined in these markets is haywire, having crashed by half in a short period in April 2006 and by two-thirds in 2008, by another 50 per cent during 2011, and yet further in 2012, thus making a mockery of the idea that there will be an effective market mechanism to make renewable energy a cost-effective investment; (6) there is serious potential for carbon markets to become an out-of-control, multi-trillion dollar speculative bubble, similar to exotic financial instruments associated with Enron's 2002 collapse (indeed, many former Enron employees populate the carbon markets); (7) as a 'false solution' to climate change, carbon trading encourages merely small, incremental shifts, and thus distracts us from a wide range of radical changes we need to make in materials extraction, production, distribution, consumption and disposal; and (8) the idea of market solutions to market failure ('externalities') is an ideology that rarely makes sense, and especially not following the world's worst-ever financial market failure, and especially not when the very idea of derivatives – a financial asset whose underlying value is several degrees removed and also subject to extreme variability – was thrown into question.[36]

---

34   FS Insight 'Global Uncertainties Denting the Carbon Market' (13 April 2012) <http://fsinsight.org/insights/detail/global-uncertainties-denting-the-carbon-market>.

35   See <http://www.durbanclimatejustice.org/>.

36   See, for example, L Lohmann 'Carbon Trading: a Critical Conversation on Climate Change, Privatisation and Power' (2006) 48 *Development Dialogue* <http://www.dhf.uu.se/pdffiler/DD2006_48_carbon_trading/carbon_trading_web_HQ.pdf>.

V   TOWARDS LITIGATIVE JUSTICE FOR ECOLOGICAL CRIMES?

What legal strategies might be deployed to evade the policy trap of carbon trading? Since 1972 when the first UN environmental conference was held in Stockholm, the war for ecological accountability has raged between the 'Global North' and 'Global South', distinctly rooted in the processes, and consequences, of imperialism and colonialism. During the 1980s and 1990s, the South fought for the UN, as a global political platform, to mediate the ecological crisis, while the North lobbied for technical bodies such as the Intergovernmental Panel on Climate Change (IPCC). The latter proclaimed the crisis as a 'universal' and technical issue, requiring regulation; while the South felt that those who caused the problem must shoulder the political and economic costs of the solution.[37] In their book, *International Relations Theory and Political Thought*, Eric Laferrier and Peter Stoett insisted that while:

> the prospect of increased co-operation on environmental issues through institutional design and growth may please many environmentalists, it may be coming as part of a package deal that in fact decreases heterogeneity, increases extractive activity, and emphasises technocratic problem solving to what are in essence political and even, philosophical dilemmas.[38]

This global condition of environmental degradation, asserts Wouter Achterberg,[39] demands that, 'the ethical basis and forms of global governance' are scrutinised, in what has become a 'singular grand narrative' of environmental ethics.[40]

As far back as the first Rio Earth Summit, the UN allocated responsibility to the primary polluters. 'The largest share of historical and current global emissions of greenhouse gases has originated in developed countries ... [and should be redressed] on the basis of equity and in accordance with their common but differentiated responsibilities,' stated the UNFCCC. Beijing's 1991 Ministerial Declaration on Environment and Development declared, 'Responsibility for the emissions of greenhouse gases should be viewed in both historical and cumulative terms, and in terms of current emissions. On the basis of equity, those countries who have contaminated most, must contribute more'.[41] Brazil's 1997 submission to the Kyoto negotiations, called for accountability on the basis of cumulative historical emissions dating from 1840.[42] Other dates range from the early 1900s to 1960s, connected to critical periods in history.

Still, the concept of ecological debt remains largely undeveloped in legal terms. It has been fundamentally opposed by the US and United Kingdom (UK), whose fossil-fuel intensive political economies remain inherently

---

37   M Friman 'Historical Responsibility in the UNFCCC' (2007) Centre for Climate Science and Policy Research Report 8(4) *Climate Policy.*

38   E Laferriere & PJ Stoett *International Relations Theory and Ecological Thought: Towards a Synthesis* (1999) 107–8.

39   W Achterberg 'Environmental Justice and Global Democracy' in B Gleeson & N Low *Governing for the Environment: Global Problems Ethics and Democracy* (2001).

40   B Gleeson & N Low 'The Challenge of Ethical Environmental Governance' in Gleeson & Low ibid.

41   A Gosseries 'Historical Emissions and Free-Riding' (2004) 11 *Ethical Perspectives.*

42   Ibid.

intertwined with the primary causes. As chief US climate negotiator Todd Stern stated at the Copenhagen COP15 in 2009, 'The sense of guilt or culpability or reparations – I just categorically reject that'. Since 1990, the US's EPA concedes domestic emissions have increased by at least ten per cent, a figure that excludes sites of pollution created by US government and corporate investment abroad, in the outsourcing of emissions-related economic activity (finished products are imported but it is China that is blamed for the emissions that go into production and transport).[43] Moreover, the lack of congressional action on climate in spite of intense pressure in the period 2007 to 2009,[44] confirms that there is no intention to reduce emissions. Conversely, the US-Canada tar-sand pipeline proves the opposite.

Most developed governments proclaim individual, and collective, innocence on the basis of the several criteria, including ignorance (if I did not know, I/we cannot be responsible); powerlessness (if I was not in charge, I/we cannot be responsible); historical emissions ignorance (if I was not alive, I/we cannot be responsible) etc. However, from a justice-based legal perspective, current and historical emissions remain a directly causal link to global and local ecological crisis. These emissions do necessitate moral, structural and material accountability where collectives or individuals (also including governments) were informed of past or current exploitation, and once informed, and having benefitted from the system, took no action to compensate.

These groups are even more accountable where the system was maintained for continuation of benefit. In terms of structural accountability for governments and companies designing energy policy, the last three decades – among the most relevant in pollutive activity, have provided ample knowledge. The question then is whether those in the political seats of power, extending to multilateral entities, did know, and did not act, due to vested interests or indifference, while engaging in free-riding or parasitical behaviour. Meanwhile, as Stephen DeCanio notes:

> In opposition, the developing countries feared institutionalization of something like current emissions levels (or ratios) that would condemn them to permanent economic inferiority because of the advantages the rich countries had derived from their historic reliance on fossil fuels to power the industrial revolution.[45]

But there is evidence of climate justice accountability being successfully pursued in US courts. Environmentally destructive activities impacting on human health have been successfully prosecuted in EU courts, as violating enforceable human rights. Moreover, reparation is a normalised and justifiable prosecutable platform. The same concept is, after all, recognised across society as a basis of 'justice in law' – hence, even the description of prisons as places where criminals, 'pay their debt to society'. The International Criminal Court,

---

43   EPA 'US Greenhouse Gas Inventory Report' (April 2012) <http://www.epa.gov/climatechange/ghgemissions/usinventoryreport.html>.

44   Climate Change Insights 'Watch what Happens: Top 5 US Climate and Energy Questions in 2012' (30 December 2011) <http://www.climatechangeinsights.com/2011/12/articles/us-policy/watch-what-happens-top-5-us-climate-energy-questions-in-2012/>.

45   S DeCanio *Economic Models of Climatic Change – A Critique* (2003).

for example, intends reparations for victims, as 'relieving the suffering and affording justice to victims not only through the conviction of the perpetrator by this Court, but also by attempting to redress the consequences ...'.[46] The case for reparations is further embedded in the institutional landscape of rights when taking into account the expanded universe of victims – via the Victim's Trust Fund – eligible for reparations as those impacted by human rights violations, rather than simply those who were victims of convicted criminals.[47] This argument steers away from the 'confrontational model', enabling an objective duty for tort reparations to be found for injured victims. This feeds into the idea of a system capable of facilitating reparations, including in financial form, rather than market mechanisms based on generating profit.

Some advocates base their hopes for such a strategy upon the GCF. The GCF is an operating entity of the financial mechanism of the UNFCCC, designed to transfer money from developed polluting countries to developing countries. The amount promised by US secretary of state, Hillary Clinton, in 2009 was US$100-billion annually starting in 2020. But in the wake of Durban's COP17 and Doha's COP18, the sources of funding were defined as from 'a wide variety of sources, public and private, bilateral and multilateral, including alternative sources of finance'. These have yet to be clarified in terms of how and when resources would be mobilised, as well as the relative share of each developed nation.

The GCF was also proposed to provide a more equal voice for developing countries in terms of the form of climate finance (whether loans or grants; for adaptation or mitigation), and potential direct access, allowing for national institutions to shape the use and outcome. But within the GCF – which has yet to identify public sources of funding – the concept of a private sector facility within the GCF, has been prioritised as a central objective for 2013, enabling 'it to directly and indirectly finance private sector mitigation and adaptation activities at the national, regional and international levels'. It is claimed that the GCF will act as a scale-up tool for private-sector funding. This bears similarity to the Bank's claims that for every US$1 invested from the Global Environmental Facility, US$4 has been attracted via co-financing.[48] Or that the Bank's private-sector lending arm, the International Finance Corporation (IFC), addresses poverty. In reality, assessments of IFC investments found less than 29 per cent flowed to the poorest countries; with almost 40 per cent to Brazil, India, Russia, China and Turkey. The Bank's own Independent Evaluation Group found that just 13 per cent of IFC investments had an 'explicit focus on the poor'.[49]

Hence as even the African Development Bank admits, while adaptation is the main concern for many low-income countries subject to extreme climate threat, mitigation received 100 per cent of private-sector funding, constituting

---

46 L Keller 'Seeking Reparations at the ICC: Victim's Reparations' (2007) <http://b6.b7.85ae.static.theplanet.com/files/faculty/Keller_reparations_ICC_final.pdf>.

47 Ibid.

48 Sharife (note 28 above).

49 Ibid.

more than 50 per cent of all climate funding.[50] Mitigation can often be a profit-rich sector insofar as vast expenditures are required to transform economies from high to low carbon. 'Offsets' were also entirely directly invested in mitigation, while just three per cent of GCF and multilateral value flowed to adaptation. In total, just five per cent of total global climate funding went to adaptation.[51]

But in assessing the evolution of international law, two themes emerge. First, reparations are primarily considered from the standpoint of a narrowly-defined 'criminal' law interpretation, and they are usually limited to deviant individuals or non-stakeholder groups. Second, reparations fail to properly consider the contradictory role of the state, which is both a primary offender and enforcer of rights. Paradoxically, the modern national state has been the cause of rights violations in the majority of cases that have occurred at mass scale. As witnessed in Nazi Germany, Australia, Canada, the US, South Africa, and other sites of systematised historic rights violation, it is frequently the state's law that was integral to the deprivation of rights. Yet despite the obvious role of the state, even the international definition of genocide fails to incorporate the state. The law, then, is not always right or wrong. While structured on legality, it does not always hold 'legitimacy' for those dispossessed by such systems, where legitimacy is, 'transcribed as a coded word for morality, thus capturing the tension between morality and legality, not the purported one between legitimacy and legality'.[52]

The essence of reparations, narrowly interpreted by international human rights norm, is constructed on five pillars: acknowledgement and apology; guarantees against repetition; measures of restitution; measures of rehabilitation; and measures of compensation. Critical components have long been normalised in international law: that states must balance the needs of victims, and the interests of the public, as well as offenders; that victims must have access to formal and informal dispute resolution mechanisms; the need for comprehensive crime prevention action; restorative justice in the form of reconciliation between offenders, victims and the public, etc. Reparations advocates underscore the essence of the system as requiring the entire community to take responsibility for righting the wrongs, where

---

50   The use of financial intermediaries (FIF) as leveraging vehicles for the private sector – many located in tax havens – remains a threat, though it is ardently backed by the EU. For example, Norway is seen as one of Africa's most vocal climate supporters. Yet, of the 35 funds invested in, 29 are based in tax havens, such as primarily Mauritius, followed by the Cayman Islands, Delaware (US), Luxembourg, and Panama. The same can be applied to others, including seven of 12 funds belonging to the European Investment Bank (EIB). Once donors disburse funds, the pledge is considered 'fulfilled' whether or not the funds reach the African country. These FIFs, operating with legal and financial opacity, are rarely monitored from the way in which funds are used since period of deposit to disbursement, nor does the jurisdiction – such as Mauritius – require any accountability. Civil society critics like Bobby Peek (groundWork South Africa) and Karen Orenstein (Friends of the Earth USA) claim that few FIFs align projects with the needs of developing countries.

51   Sharife (note 28 above).

52   J Connor 'The Meaning of "Legitimacy" in World Affairs: Does Law + Ethics + Politics = A Just Pragmatism or Mere Politics?' (2007) SGIR conference presentation <http://www.turin.sgir.eu/uploads/oconnor-legitimacy.pdf>.

society has benefitted, and possibly continues to benefit. In the case of African Americans, society would be classified as the wrongdoer, particularly where, 'the injured party is still injured and suffers from the consequences of the wrong'.[53] In this instance, reparations for damages (the basis of tort) would necessarily require restorative justice.

The impact of policy choices (whether at a national or supranational foreign policy level concerning oil partners, or at the domestic level of waste management) is directly mirrored in the political ecology (defined in this context as politically regulated engagement between natural and human ecologies). Though scientific commissions such as the IPCC openly bridge invisibilised gaps between ecological degradation and human rights, little has been done to move towards reparations and restorative justice.

Gains on the legal front do suggest reparations will be increasingly litigable, yet the litigative landscape is too often interpreted via neo-liberal politico-economic values. Outcomes are too often aligned to facilitate corporate penetration of markets through commodifying as much as possible, with minimal state regulation, intervention or protection. The market considers only profit, with all other factors diminished to externalities. Still, a few recent cases indicate patterns of legal justice may possibly provide standing against blind commodification.

For example, in the landmark case, *Friends of the Earth (FoE) Inc v Spenelli*, FoE and four US cities (Boulder, Oakland, Santa Monica and Arcata) filed a lawsuit against two US agencies, the Overseas Private Investment Corporation (OPIC) and the Export-Import Bank (Ex-Im), alleging that the US\$32-billion provided for financial and political risk insurance to fossil-fuel intensive projects, did not consider the heavy impact on climate change. Between 1990 and 2003, the plaintiffs alleged the OPIC and Ex-Im funding generated eight per cent of global GHG emissions (including 260-million CO_2 annually). Though the case was settled out of court in 2009, under the Obama administration, the 9th Circuit ruling that the plaintiffs had legal 'standing' constructed an important precedent for future related climate change actions.[54] Despite the US government's staunch opposition, it was the first lawsuit by private or public plaintiffs to hold the US government to account for its contributions to climate change, damaging the global environment and the US citizenry.[55]

While the outcome of the settlement has been anything but equitable – Ex-Im, for instance, has reverted to business as usual, including supporting the more rapid extraction of South African coal via US\$1-billion worth of Wisconsin-made machinery – the precedent of legal standing was critical. Until this case, citizen-suit provisions enabling citizens etc to act as private attorneys in protection of the environment had been blocked by a series of legal federal and national court rulings. The heyday for legally-grounded environmental

---

53   Quotation by Vincene Verdun. For more see V Verdun 'If the Shoe Fits, Wear It: an Analysis of Reparations to African Americans' (1993) 67 *Tulane LR* 597–698, 612–639.
54   See Ex-Im Settlement Agreement *Friends of the Earth Inc v Spenelli*.
55   C Carlane *Climate Change Law and Policy: EU and US Approaches* (2010).

challenges in the US was the 1980s, as lawsuits increased from 41 in 1982 to 266 in 1986, with overwhelmingly successful results: 503 of 507 private suits were won by environmental claimants by the late 1980s. The demand for injuctive relief and legal costs differs from civil lawsuits, where plaintiffs are entitled to seek damages personally.

But then, as the *New York Times* reported in 1999, the success of private suits was sharply cut down following a series of Supreme Court decisions by Justice Antonin Scalia, beginning with a *Law Review* article authored in 1983, three years before he was appointed to the Supreme Court, designed to prevent prosecutable civil suits. When asked whether his intention was to 'misdirect' and erode environmental citizen suits, Scalia responded 'of course ... and a good thing too'.[56] Even more broadly considered, rights-based approach through courts often operate in severely constrained spaces,[57] where rights are exercised according to the interpretation of hegemonic systems, excluding intergenerational justice[58] and equity,[59] such as state-capitalism that brooks no challenges (especially China), the US's neo-liberal system, or the EU's capitalist-welfare model.

In spite of the fact that universally-recognised 'human dignity' is enshrined in multiple global, regional and national constitutions, conventions, treaties etc, from the UN to the EU and below, there is still very infirm ground to stand upon for legal precedent in protecting the environment and demanding reparations for damages. Allocation and protection of equal basic rights, including the distribution of scarce resources, should ideally be adjudicated through a globally-determined framework, and be mandatory. Yet, even though more than 100 inalienable human rights treaties exist, the legal application and operationalization of these instruments has not yet been clarified, nor is there significant movement to that end.[60]

Nevertheless, numerous climate-related lawsuits continue, in relation to corporate torts against people in Nigeria, the US, South Africa, Canada, Australia, New Zealand etc. The lawsuits typically focus on curbing the excesses of corporate choice, and their resulting human and ecological

---

56   W Glaberson 'Novel Antipollution Tool is Being Upset by Courts' *New York Times* (5 June 1999) <http://www.nytimes.com/1999/06/05/us/novel-antipollution-tool-is-being-upset-by-courts. html?pagewanted=all&src=pm>.

57   H Osofsky 'The Geography of Climate Change Litigation: Implications for Transnational Regulatory Governance' (2005) 83 *Washington Univ LQ* 1789, 2005 <http://ssrn.com/ abstract=796204>.

58   Justice being interpreted here as an inviolable, universal moral compass, perceived as Aristotle stated, 'justice alone, of all virtues, is thought to be "another's good" because it is related to our neighbours; for what it does is advantageous to another ...'

59   Equity in this context referring to various justice-based contexts, ie interspecies, intergenerational etc, defined as 'forms of justice that remedies the injustice of ... law at the point of application by attending to particular features of the persons and circumstances involved ... so that equity is not generically different from justice ... (J Tasioulas 'Justice, Equity and Law' in E Craig (ed) Routledge Encyclopaedia of Philosophy (1989) 148).

60   The ICJ has, via cases such as Barcelona Traction Light & Power Co (*Belgium v Spain*) 1970 ICJ 3, 32 (5 February judgment); Military and Paramilitary Activities (*Nicaragua v US*) 1986 ICJ 14, 114 (27 June judgment) identified that UN member states have inviolable human rights obligations, but save for exceptions, the scope of the rights has yet to unpacked and implemented due to opposition from political blocks of power.

'externalities' – or hidden costs. The Inuit petition, filed by the Center for Environmental Law (CIEL) in 2005, with the Inter-American Commission on Human Rights (IACHR), identified a direct binding obligation on states by international rights law. The petition noted that the US, with five per cent of the population, generated 25 per cent of global GHG emissions, and not only refused to participate in negotiations for genuine reductions, but further, 'actively impeded the ability of the global community to take action'.[61] In a supplemental testimony to that provided by Sheila Watt-Cloutier,[62] who described indigenous peoples sustainably inhabiting ecologies, as 'early warning systems', attorney Martin Wagner identified the impacts of global warming as violating three specific sets of rights: the rights of indigenous people; the right to life, security, and physical integrity; and the right to use property without undue interference.[63]

The 'right to life' for instance, is defined by international scholars as holding, 'access to the means of survival, realize full life expectancy, avoid serious environmental risks to life, and to enjoy protection by the State against unwarranted deprivations of life'.[64] Customary international law obligates every state 'not to allow knowingly its territory to be used for acts contrary to the rights of other States'.[65] Yet while giving the matter time and attention, the IACHR rejected the petition, providing only a hearing. This was similar to *Massachusetts v EPA*,[66] the first climate change case[67] to reach the US Supreme Court. The suit was driven by a coalition (including 12 states such as California, Washington, Oregon; cities such as Washington and NY; environmental organisations etc) who petitioned the EPA to regulate GHG emissions under the Clean Air Act s 202(a)(1), harming public health and welfare, such as the coast of Massachusetts. The EPA denied the request, claiming that even if Massachusetts was aided, to combat global warming, the whole of the US would have to be regulated. The Court ruled 5:4, upholding the EPA's stance.

Once again, Scalia was instrumental, stating that the 'statute says nothing at all about the reasons for which the Administrator may defer making a judgment ...' and that courts should not get involved in EPA matters, as the EPA is the specialist authority. Yet the plaintiffs were in the right: for a suit to have legal standing, legal pillars such as injury, causation and redressability are required, all of which was sufficiently proved. Scalia's views on reparations

---

61 Carlane (note 55 above).
62 Backed by the Inuit Circumpolar Conference, an NGO representing a constituency of 150,000 Inuit based in Canada, Alaska, Greenland, and Russia, filed on 'behalf of all Inuit of the Arctic regions of the United States and Canada'.
63 M Wagner 'Testimony of Earthjustice Managing Attorney Martin Wagner before the Inter-American Commission on Human Rights, 1 March 2007' (2010) cited in Carlane (note 55 above).
64 B Ramcharan 'The Concept and Dimensions of the Right to Life' in *The Right to Life in International Law* (1985).
65 Wagner. Cited in Carlane (note 55 above).
66 *Massachusetts v Environmental Protection Agency* 549 US 497 (2007).
67 For a summary of some key climate change cases worldwide, see S Stone & D Grossman 'Summary of Climate Change Cases' INECE <http://www.inece.org/conference/7/vol2/33_ClimateChange.pdf>.

for state violations of human rights is well known: 'Individuals who have been wronged by unlawful racial discrimination should be made whole; but under our Constitution there can be no such thing as either a creditor or a debtor race'.[68] When engaging the issue of plaintiffs seeking justice via the Alien Tort Claims Act (ATCA) – designed for this very purpose, Scalia stated, '... Some matters ... are none of its business', but this did not include Scalia weighing down heavily in *United States v Eichman*, when he declared the burning of the American flag unconstitutional.[69]

The ATCA itself, the most crucial instrument for reparations in the US, may soon be drastically trimmed down: currently, the ATCA allows US courts jurisdiction over claims from foreigners for international law violations, specifically human rights.[70] This usually takes the form of a confrontational model between victims who allege violations, and the corporations responsible. This may not be as dire to activists as might appear: the case brought forward by Nigerian political activists against Shell for aiding and abetting a dictatorship, is a case in point. In their argument, Shell would claim[71] that corporate liability under international law does not exist. In June 2009, Shell agreed to an out-of-court settlement with reparations payments of US$15.5-million. Representing just four hours' worth of Shell profits, it was considered by some a crucial step in establishing liability and disincentivising corporate exploitation of people and nature. But others keenly note that Shell won a massive victory by avoiding trial and 'closing' the case, thus avoiding a court undressing of Shell's corporate affairs. The lawsuit represented window-dressing symbolism in this reading. The context, despite Justice Samuel Alito's contrary statement, is that Nigeria is a key oil supplier to the US, integral to Washington's foreign policy and energy security. But in his statement querying whether the case really belonged in the US, conferring a kind of 'legal Rome' status, Alito rightly probed whether international judicial systems might be a better option.

Beyond these kinds of tort actions, will courts start declaring climate-related ecological debt a valid concept? The Ecuadoran court system is one site for exploring constitutional rights to nature, and Bolivia may follow. In February 2011, a US$8.2-billion judgment against Texaco was won by Ecuadorans in local courts due to decades of Amazon oil pollution, which then doubled because of the failure of the firm's owner Chevron to immediately apologise. Chevron refused to recognise the judgment and filed a countersuit of racketeering against the plaintiffs' lawyers.[72] An international climate

---

68   A Scalia Concurring Opinion *Adarand Constructors Inc v Pena* (No 93-1841). See <http://www.law.cornell.edu/supct/html/historics/USSC_CR_0515_0200_ZC.html>

69   Rescued from dormancy in 1980 when the 2nd US Circuit Court of Appeals ruled that US courts had jurisdiction over international rights violation in *Filartiga v Pena-Irala*, 630 F 2d 876 (1980).

70   Other critical ATCA cases include reparations for foreign multinationals buttressing the apartheid regime in South Africa, the infamous Chiquita fruit farms in Colombia, Exxon and Chevron operations in Indonesia etc.

71   Shell as respondents, see <http://www.losangelesemploymentlawyer.com/International-Human-Rights/Supplemental-Brief-for-Respondents-8-1-12.pdf>.

72   D Edwards 'Chevron Sues Ecuadorians who Stood up to Toxic Contamination' (2011) *The Raw Story* <http://www.rawstory.com/rs/2011/02/14/chevron-sues-ecuadorians -who-stood-up -to-toxic-contamination/>.

tribunal is one of the climate justice movement's objectives, similar to the International Criminal Court in The Hague. Although at the time of writing the final judgment had not been made, oil company BP was fined in the region of US$20-billion for its Deepwater Horizon spill in the Gulf of Mexico.

Indeed the power of reparations, at its core, is a moral claim located in a landscape of politics. But, as we have seen, when placed against the backdrop of rights-bearer, interpreted by the legal arms of political powers that have benefitted from the injury, transnational legal routes for litigating rights-talk are unlikely to provide justice for injured human and natural ecologies. In identifying four major preconditions for successful ecological regimes, Ian H Rowlands highlights the importance of fairness, 'in a system wide sense ... including less tangible factors such as history and ideology ...'.[73] Indeed, for reparations and restorative justice to be realised, an equitable politics separate from lethal market-mechanisms and capitalist ideologies must be devised. This must factor in both losses suffered by the victims, and gains accrued by the beneficiaries, distributed collectively to collective masses, prioritising interspecies equity and reparations – the most vulnerable, the most irreplaceable, of all victims. These decisions must be realised as supra-sovereign and mandatory, political decisions. Where monetary compensation is necessary as one complementary strategy, funds should be distributed outside of rent-seeking state patronage machines, through systems such as Namibia's pilot version BIG.[74]

Yet, ecological debt advocates must be very careful not to adopt the capitalist conception of justice-served by diminishing damage to monetary compensation only. The self-defeating, narrow and economistic nature of legal torts, realised through money, is infected with the pitfalls of capital as the sole compensation, *via personal reparations* as the goal. For reparations to be just, collective restorative action must underpin political imperatives at a state and trans-state level, rather than primarily pursued through courts. What differentiates the two, among other fundamental distinctions, is that collective reparations involve accountability that must be evinced from *a change in structural and systemic architecture*, as opposed to merely a 'fine' for wrongdoing. In other words, the 'fine' (polluter pays) must be followed by a *ban* on further pollution, for there to be restorative justice. Theories of justice must not be limited to financial compensation as the core strategy, or delinked from the root cause of injustice, ie that the state as a sovereign power influenced by powerful for-profit institutions, determines how pollution is created, by their own policies, and by those of their legal and human citizens, including companies like Shell. At the heart of climate change, of course lies the ability of states to 'write laws' facilitating legal and financial environments of opacity, extending from human

---

73   IH Rowlands 'Ozone Layer Depletion and Global Warming: New Sources of Environmental Disputes' (1991) 16 *Peace and Change* 260.

74   Investigative research by Ejolt-CCS revealed that funds remitted via the Nampost system were corruption free. As citizens cannot often, if ever, hold governments to account, avoiding this crucial fault line is necessary. Africa's primary source of maldevelopment is derived from capital flight, 60 per cent caused by corporate mispricing, estimated at US$200-billion minimum annually.

rights to environmental regulation. The Netherlands, for instance, where Shell is based, hosts over 355 dirty industry subsidiaries. The same country provided a sovereign loophole as a tax haven for mega-multinationals sustaining the apartheid regime. Justice for the climate would thus require justice at various levels, including the natural sanctioning of secrecy jurisdictions.

In this context, imposing a fee for ecosystem services and other pricing strategies leading to environmental markets must be abandoned. Instead, prohibitive fines are needed to incentivize preventative action; bans must be implemented for certain activities, such as gas flaring; and reparations in the form of ecological debt must finally begin. The various past and future climate-regime meetings can either accept this reality, or continue privatising nature, with all the market failures we have come to expect. And if this continues, yet more litigation for various forms of ecological debt is inevitable.

VI   CONCLUSION: HOW, THEN, TO PAY THE ECOLOGICAL DEBT?

At the end of this process, there necessarily arises the question of how, if either legal or direct action pressure permits climate debt to become part of Northern climate concessions, the payments are best distributed. It became clear to many civil society groups in recent decades that post-colonial African governments were too easily corrupted, just as were UN agencies and aid (and even international NGO) bureaucracies. As one reflection, the late Ethiopian tyrant Meles Zenawi – Africa's leading climate debt advocate, ironically – in July 2011 announced the purchase of 200 tanks from Ukraine for over US$100-million even while climate change caused the worst drought in 60 years, leaving millions of his subjects hungry. When demands for democracy and human rights soon emerged, he jailed more opposition party leaders and expelled Amnesty International, and was caught by the BBC using development aid as a reward for impoverished citizens' political loyalty, penalising non-members of his party by denying them food.[75]

The solution to the complementary payment distribution problem appeared in 2009: the idea of simply passing along a monthly grant – universal in amount and access, with no means-testing or other qualifications – to each African citizen via an individual BIG payment. According to *Der Spiegel* correspondent Dialika Krahe, the village of Otjivero, Namibia, was an exceptionally successful BIG pilot for this form of redistribution:

> It sounds like a communist utopia, but a basic income program pioneered by German aid workers has helped alleviate poverty in a Nambian village. Crime is down and children can finally attend school. Only the local white farmers are unhappy ... 'This country is a time bomb,' says Dirk Haarmann, reaching for his black laptop. 'There is no time to lose. Haarmann and his wife Claudia, both of them economists and theologians from western Germany, were the ones who calculated the basic income for Namibia. 'The

75   A Ashine 'Ethiopia Expels Amnesty Group and Jails Opposition Party Leaders' *Uganda Monitor* (2 September 2011) <http://www.monitor.co.ug/News/World/-/688340/1229108/-/11hh59e/-/?utm_medium=twitter&utm_source=twitterfeed>; BBC 'Ethiopia: Using Aid as a Weapon of Oppression' (25 August 2011); T Mountain 'Ethiopia Buys Arms as Millions Starve' *Counterpunch* (7 July 2011).

basic income scheme,' says Haarmann, 'doesn't work like charity, but like a constitutional right'. Under the plan, every citizen, rich or poor, would be entitled to it starting at birth.[76]

According to leading African investigative journalist John Grobler, within the BIG system – allocating R100 per month for those under 60 years old, a sum paid out for two years, between January 2007 and December 2008 – 'there is no space for corruption in the system'. The impact on social indicators, including food, education and health security, dramatically increased. Contrary to opposition within government, no new shebeens were opened – instead, shebeens were closed on BIG payout days. Meanwhile, domestic violence, begging and sex work (and with it, transfer of HIV and STDs) declined considerably.

If this strategy was to be generalised, the first priority would be to introduce a BIG programme to low-income people who live in areas most adversely affected by droughts, floods or other extreme weather events. Logistically, the use of Post Office Savings Banks or rapidly introduced Automated Teller Machines (ATMs) and specialised withdrawal accounts would be sensible, although currency distortions, security and other such challenges would differ from place to place. How has BIG or similar income transfer schemes worked elsewhere? Ironically, in oil-dependent, conservative Alaska, the BIG concept – in the form of a dividend from oil-sourced income that goes to every Alaskan citizen – has been welcomed across the ideological spectrum. Does such a grant to working-class people disincentivise them from seeking work? According to Guy Standing of the International Labour Organisation, the BIG can actually generate 'a greater incentive to search and to take jobs, particularly low-wage jobs or low-income, own-account activities'.[77] The reason is that BIG is universal and does not get withdrawn once a job is taken. (On the other hand, the BIG can, of course, be used against labour, as much as for labour, as Franco Barchiesi records; and as James Ferguson notes, BIG is a social policy that can readily operate within the logic of neo-liberalism.[78])

In short, when it comes to monetized reparations for loss and damage from climate change, a modified version of an income grant, specific to the impact and geographical area affected, would provide due legal compensation to individuals affected. Still, once that has been accomplished, a much deeper strategy for financing collective climate adaptation or survival strategies – not to mention full transformation of socio-economic systems to achieve a post-carbon world – would still be required. But the question still arises: is monetized accounting appropriate within the concept of ecological debt, which is described by Martinez-Alier as a debt owed also to future generations and species.

As Karin Mickelson notes:

---

76   D Krahe 'A New Approach to Aid: How a Basic Income Program saved a Namibian Village' *Der Spiegel International* (10 August 2009).

77   G Standing 'The South African Solidarity Grant' in G Standing & M Samson (eds) *A Basic Income Grant for South Africa* (2003).

78   F Barchiesi 'Liberation of, through, or from Work? Postcolonial Africa and the Problem with "Job Creation" in the Global Crisis' (2012) 4 *Interface: a Journal for and about Social Movements* 230; J Ferguson 'Formalities of Poverty: Thinking about Social Assistance in Neoliberal South Africa' (2007) 50(2) *African Studies Review* 71.

ecological debt traces many of the benefits presently enjoyed by the North to its longstanding ability
to draw upon global resources. The moral responsibility with regard to ecological debt, in other words,
does not derive from visiting the 'sins of the fathers' upon present generations in the North. Instead, it
arises from an acknowledgment that the privileged position of the developed countries represents the
culmination, and in many cases the perpetuation, of a history of unequal access.[79]

All of this is still rejected by the North. 'The concepts of "ecological costs" and
"ecological debt" are highly controversial' for European elites, according to
Jos Delbeke, the European Commission's director-general for Climate Action.
'The concept of "North" versus "South" does no longer hold. Therefore, as
agreed at last year's Durban conference, we need a new legally-binding global
agreement covering all countries – by 2015 at the latest.'[80]

But given the ability of Washington – joined by BRICS [Brazil, Russia,
India, China, South Africa] countries – to sabotage binding cuts in the
UNFCCC, the multilateral negotiations now appear permanently doomed.
The burning problem of climate debt can otherwise be addressed through
solidaristic initiatives – in the way German Lutherans responded to the
Namibian BIG opportunity[81] – and through litigation, such as is seen in
embryonic form in the ecological debt cases discussed above. In this way,
a strategy emerges with much greater ambitions: to ensure that the GHG
'polluters pay' in a manner that first, compensates their climate change
victims; that second, permits transformation of African energy, transport,
extraction, production, distribution, consumption and disposal systems; and
that third, in the process assures the 'right to development' for Africa in a
future world economy constrained by emissions caps. Extremely radical
changes will be required in all these activities in order not only to ensure the
safety of the species and planet, but also to confirm that Africans are at the
front of the queue for long-overdue ecological and economic compensation,
given the North's direct role in Africa's environmental damage.

Indeed the contemporary argument for climate debt to be paid is simply the
first step in a long process, akin to decolonisation, in which the master – the
polluting global North (the wealthy of South Africa included) – must know
that the tools with which his house were built, including the profit motive
and markets, are not – and cannot be – effective in remodelling for a new
society. And he must also know that not only is it time to halt the reliance
on fossil-fuels, but having 'broken' the climate, it is his responsibility to foot
the clean-up bill. And that work of valuation and pricing, finally, must be
done *without creating new market strategies to privatise nature* and to worsen
world financial chaos and corruption. The climate debt must be implemented
so as to begin to achieve justice, via direct payments to the victims of the
crisis the master has created.

79   K Mickelson <http://faculty.law.ubc.ca/hsu/climate_seminar/GORD_REV_readings1.pdf> 6.
80   Cited in Sharife (note 28 above).
81   For an exemplary form of such solidaristic popular education, see the film 'The Bill' produced
     by eco-film <http://www.eco-film.de/-> and sponsored by GermanWatch <http://germanwatch.
     org/klima/film09e.htm>.

CLIMATE JUSTICE IN COURT

LET'S WORK TOGETHER: ENVIRONMENTAL AND SOCIO-ECONOMIC RIGHTS IN THE COURTS

JACKIE DUGARD AND ANNA ALCARO[1]

> Environmental concerns can unite South Africans going beyond economic and political barriers.
>
> Nelson Mandela, 15 August 1993[2]

I INTRODUCTION

As some of the chapters in this book emphasise, climate change increasingly threatens natural and social systems, in many instances placing strains on relationships and apparently pitting one system against another. For example, as fears mount regarding global water scarcity, the right to water for human domestic consumption is increasingly being contested in relation to water resource conservation.[3] Yet climate justice demands even closer links between systems. So, how does this paradox play out in South Africa with its troubled history of de-linking among social systems and between these and natural systems and how, if at all, has a post-apartheid rights-based paradigm assisted attempts to pursue a more integrated approach?

Acknowledging South Africa's fragmented past, the Constitution of the Republic of South Africa (1996) contains a set of synergistic environmental rights that can be seen to combine natural and social perspectives of the

1 The authors are grateful to the Chr Michelsen Institute in Bergen (Norway) for partnering with the Socio-Economic Rights Institute of South Africa (SERI) to undertake a broader project on climate change, the environment and poor communities for which this chapter was an output. We are also grateful for useful comments from Tracy-Lynn Humby, Rachel Wynberg and Michael Clark, as well as two anonymous reviewers. And we would like to extend our appreciation to Rose Williams (Biowatch), Desmond D'sa (SDCEA), and Samson Mokoena (VEJA) who – with generosity and patience – offered their insight into specific struggles for environmental justice. Finally, we are indebted to the communities and activists of South Durban, including Bobby Peek, Vanessa Black, Bongani Mthembu, Nawaal Domingo, Abraham Mei, Delwyn Pillay and Patrick Bond.

2 International Development Research Centre (IDRC) 'Building a New South Africa: Environment, Reconstruction and Development' 4 (1995) cited on inside cover <http://idl-bnc.idrc.ca/dspace/bitstream/10625/15055/25/102795.pdf>.

3 In the South African case of *Mazibuko v City of Johannesburg* 2010 (4) SA 1 (CC), the government's rejection of a request to increase the provision of free basic water to poor households was underwritten by arguments about the country's overall water scarcity, which required 'careful management' (para 3 of the *Mazibuko* judgment). Yet, this pitting of social against natural systems ignores the fact that domestic water consumption – and particularly that by poor households – uses only a fraction of the country's water reserves (around 10 to 15 per cent), with agriculture and industry using the bulk of South Africa's water supply. The artificial juxtaposition also disregards environmental justice advocates' concerns about the commodification of water and the impact of this on social and environmental systems (T-L Humby 'Environmental Justice and Human Rights on the Mining Wasteland of the Witwatersrand Gold Fields' unpublished paper).

environment.[4] Thus, s 24 of the Constitution stipulates that conservation should be promoted and pollution and degradation prevented, and also guarantees everyone's right to an environment that is not harmful to their health or well-being and that supports justifiable economic and social development. The Constitution also entrenches a range of socio-economic rights, including rights of access to adequate housing (s 26) and to sufficient food and water (s 27). Given the apartheid legacy of environmental and human degradation, as well as the characterisation of the environment as being a natural system for the benefit of the white minority, adopting an integrated environmental rights approach that incorporates social systems and socio-economic rights is essential.[5] Indeed, from a socio-economic rights perspective, it is clear that 'rights to food, healthcare and water[6] are inextricably woven together with the realisation of environmental rights'.[7]

Yet, few studies have sought to explore the relationship between environmental and socio-economic rights,[8] particularly from a legal angle, and South African environmental case law overwhelmingly reflects a narrow definition of the environment that is not infused with social justice issues. This lacuna has prompted Loretta Feris to call for the development of jurisprudence aimed at 'clarifying the relationship between s 24 and other human rights'.[9] As pointed out by Feris and others, in perhaps the only case in which the Constitutional Court has had to explicitly grapple with environmental and socio-economic rights – Kyalami,[10] in which the Court was asked by the Kyalami Ridge Environmental Association to overturn, on environmental (and other) grounds, a decision by the government to provide temporary shelter for victims of flooding – the Court presented housing and environment as conflicting rights, with environmental rights applying to the richer property owners and housing rights applying to the poorer flood victims.[11] While this bifurcated, zero-sum gain approach is perhaps to be expected when environmental rights are deployed in defence of property values, as was the case in Kyalami (or business ventures as has occurred in other litigation) and detailed below, it appears from the case law that this is also so when environmental movements and even environmental justice movements litigate on environmental issues.

---

4   L Feris 'Constitutional Environmental Rights: An Under-Utilised Resource' (2008) 24 *SAJHR* 29–49.
5   R Wynberg & D Fig 'Realising Environmental Rights: Civil Action, Leverage Rights, and Litigation' in M Langford, B Cousins, J Dugard & T Madlingozi (eds) *Symbols or Substance: The Role and Impact of Socio-Economic Rights Strategies in South Africa* (Forthcoming 2013) 5.
6   Constitution s 27.
7   Wynberg & Fig (note 5 above) 7.
8   Ibid 1.
9   Feris (note 4 above) 38.
10   *Minister of Public Works v Kyalami Ridge Environmental Association* 2001 (3) SA 1151 (CC).
11   Wynberg & Fig (note 5 above) 1. While most commentators have applauded the Court for upholding the government's decision to act positively in terms of its housing-related obligations, the judgment has been criticised for posing environmental rights and housing rights as necessarily oppositional (see for example Feris (note 4 above) 42–3.)

But is this necessarily the case, or is there a way to pursue a more integrated approach that combines environmental ('green'[12]) and social justice ('brown'[13] and 'red'[14]) issues? This chapter engages with such questions by first examining the kinds of issues taken up and mobilisation that emanates from some of the nascent environmental justice movements in South Africa. Focusing on the South Durban Community Environmental Alliance (SDCEA) (pronounced sĕd-sēa), the chapter outlines the contours of emerging environmental justice mobilisation before examining the uptake of litigation by SDCEA and other similar environmental justice movements. Turning our lens to the courts, the chapter then analyses the limits and opportunities of environmental litigation to date, concluding that, in a series of similar cases launched in mid-2012, there might be the beginnings of a more integrated approach to environmental litigation.

II ENVIRONMENTAL JUSTICE AND RIGHTS-BASED MOBILISATION

The relationship between rights and social mobilisation has attracted significant scholarship, including that of Michael McCann who, in the context of gender pay equity reform in the United States of America (US), noted in his much-quoted book, *Pay Equity Reform and the Politics of Legal Mobilisation*,[15] on how rights became a mechanism for achieving social change.[16] On the ground, social movements around the world have been using rights – with or without litigation – as part of their mobilising arsenal for some time. Post-apartheid South Africa is no exception and, with an expansive Constitution, mobilised activism has often involved litigation in what has been referred to as the 'legal-activist' approach.[17] In the well-known case of the Treatment Action Campaign (TAC), activists mobilised around the right to health and mounted a successful legal challenge against the government's policy of restricting to pilot sites the provision of antiretroviral medication to pregnant mothers (to prevent mother-to-child transmission of HIV/AIDS). More recently, the shack-dwellers' movement, Abahlali baseMjondolo, has begun to utilise rights as

---

12 While there is an overall understanding that green issues concern plant and animal species, as well as the natural environment and therefore pollution, there is a spectrum of green ideologies ranging from 'dry greens', who believe that the market and benign self-regulation can address such concerns, to 'deep greens', who reject consumerism and the capitalist project (J Cock 'Connecting the Red, Brown and Green: The Environmental Justice Movement in South Africa' (2004) *Globalisation, Marginalisation & New Social Movements in Post-Apartheid South Africa* (unpublished paper) 2 <http://www.swopinstitute.org.za/node/21>).

13 Brown issues relate to human well-being especially in the context of urban settlements, and usually include basic services such as electricity, water, sanitation and waste removal.

14 Red issues relate to human security and typically include housing, social security, health care and employment.

15 M McCann *Rights at Work: Pay Equity Reform and the Politics of Legal Mobilisation* (1994).

16 Examples presented by (McCann Ibid; S Scheingold *The Politics of Rights* (1974); C Geertz 'Ideology as a Cultural System' in D Apter (ed) *Ideology and Discontent* (1964) 47).

17 R Greenstein 'State, Civil Society and the Reconfiguration of Power in Post-Apartheid South Africa' (2003) *Centre for Civil Society (CCS) Research Report* 8 <http://ccs.ukzn.ac.za/files/Greenstein%20report%208.pdf>.

part of its organising and mobilising strategy and won a significant housing rights-related case in the Constitutional Court.[18]

Environmental movements have been no exception. And, although mainstream environmental activism in South Africa still predominantly focuses on green issues (and related concerns, sometimes including pollution) such as those pursued by, for example, the Endangered Wildlife Trust, there is an emerging environmental justice movement, which amalgamates the ecological and social justice issues affecting the most poor and marginalised people in society. After all, under apartheid spatial planning, black townships bore the brunt of pollution as dirty industry was deliberately placed next to them exposing vulnerable populations to various health risks.[19]

Writing about the environmental struggles of the early-2000s, Jacklyn Cock identifies a new patterning of environmental grassroots mobilisation seeking to involve a mix of social justice and environmental issues, and fuelled by the frustration over the 'crisis experienced by poor, vulnerable communities without access to jobs, housing, land, clean water, and sanitation'.[20] At the time of writing (2004), Cock cautioned that it was still unclear the extent to which constitutional rights frameworks could successfully provide a common voice of demands for those involved, but she optimistically noted:

> This embryonic environmental justice movement is bridging ecological and social justice issues in that it puts the needs and rights of the poor, the excluded and the marginalised at the centre of its concerns. It is located at the confluence of three of our greatest challenges: the struggle against racism, the struggle against poverty and inequality and the struggle to protect the environment, as the natural resource base on which all economic activity depends. The movement is stratified in a complex layering involving national networks, NGOs and local grassroots groups. Within this multiplicity of organisational form, the vitality of the movement flows from the bottom up, being driven by the unemployed and lower working class, the 'poors'. This social base is distinctively different from the middle class composition of the mainstream environmental movement, which focuses on curbing species loss and habitat destruction, that is on 'green' issues.[21]

Under this rubric, environmental justice is 'an all-encompassing notion that affirms the use value of life, all forms of life, against the interests of wealth, power and technology'.[22] It involves 'empowered people in relations of solidarity and equity with each other and in non-degrading and positive relationships with their environments'.[23] Whereas in the US, the environmental justice movement has focused on environmental racism and the symptoms of environmental degradation, in South Africa, the evolving environmental justice movement has focused on class issues and the root causes of

---

18 *Abahlali baseMjondolo Movement SA v Premier of the Province of KwaZulu-Natal* 2010 (2) BCLR 99 (CC). In this case, the Constitutional Court declared unlawful the operative section of the KwaZulu-Natal Slums Act, thereby effectively burying the problematic legislation that allowed the province to evict residents of informal settlements without consultation.

19 Cock (note 12 above) 4.

20 Ibid 30.

21 Ibid 1–2.

22 M Castells *The Power of Identity: The Information Age: Economy, Society and Culture* (1997) 132.

23 D Hallowes & M Butler 'The Balance of Rights – Constitutional Promises and Struggles for Environmental Justice' (2004) *The groundWork Report* <http://ebookbrowse.com/the-balance-of-rights-2004-groundwork-report-pdf-d68849372>.

environmental degradation – such as deregulation and privatisation. For Cock, '[in South Africa] environmental justice means social transformation directed to meeting basic human needs and rights'.[24] In this context, environmental rights are tools that can be used to legitimise claims and challenge power relations vis-à-vis government, mining, business or industry, or private power-holders; and whether in the form of confrontation, engagement or litigation. And environmental justice is a platform that holds the potential to collectively mobilise and vindicate unsatisfied social needs.

One of the first South African movements to embody this evolving form of environmental rights in South Africa was the Environmental Justice Networking Forum (EJNF). It was launched as a network in 1994 and was intended to 'bring together a rainbow alliance of the rainbow nation'.[25] EJNF sought to bring together a cross-class and race alliance committed to social transformation through an expanded definition of environmental justice – according to the EJNF Charter:

> Environmental justice is about social transformation directed towards meeting basic human needs and enhancing our quality of life – economic quality, health care, housing, human rights, environmental protection, and democracy. In linking environmental and social justice issues the environmental justice approach seeks to challenge the abuse of power which results in poor people having to suffer the effects of environmental damage caused by the greed of others.[26]

Signalling a decisive break with the narrow, elite conservationism under apartheid, EJNF founder, Chris Albertyn, underscored that 'the environment is not just areas of natural beauty but is in fact the place where we are – the environment includes our workplace, home, hostel, town, village and city as well as areas of natural beauty'.[27] However, as documented by Cock, holding together such diverse interests was difficult and there were also enduring problems in the relations between the various participating organisations (at its height, there were over 400 participating organisations) and the head office.[28] Ultimately, EJNF collapsed in 2005 and was replaced 'by a multiplicity of activism in various arenas'.[29]

Another one of the new-generation environmental justice movements in South Africa is the Vaal Environmental Justice Alliance (VEJA),[30] which grew up out of the (mostly unsuccessful) battles waged during the late-1990s by the residents of Steel Valley (Vaal triangle, south of Johannesburg) against ISCOR[31] (discussed in part IV). Located in the Vaal triangle, which is the hub of South Africa's chemical and steel industry and home to substantial mining and smelting operations, VEJA was established in 2005. Today VEJA comprises approximately 15 different environmental, community and labour

---

24   Cock (note 12 above) 6.
25   D Hallowes (1994) 3 *The Networker* 3.
26   'Environmental Justice Networking Forum (EJNF) Charter' cited in Cock (note 12 above) 7.
27   C Albertyn 'Towards Sustainable Reconstruction' (1995) 1 *Environmental Justice Networker* 1, 9.
28   Cock (note 12 above) 8.
29   Wynberg & Fig (note 5 above) 9.
30   See <http://vaalenvironmentalnews.blogspot.com/2011/08/what-is-vaal-environmental-justice.html>.
31   Later sold to Mittal Steel, which subsequently became known as Arcelor Mittal.

organisations whose aims are to address the social, ecological, political and economic problems in the area. In particular, VEJA attempts to tackle the harm created by the big chemical and steel companies in the area, including pollution and corruption. VEJA promotes the rights in South Africa's Constitution and has begun to use litigation, but to date this has been mainly in the narrow form of access to information requests for documents from various industrial players using the Promotion of Access to Information Act 2 of 2000 (PAIA). Similarly, as discussed in part IV, Earthlife Africa, which is a voluntary environmental justice movement that was founded as far back as 1988 and today acts as an environmental justice advocacy group, has litigated on narrow access to information and review of administrative decisions using PAIA and the Promotion of Administration of Justice Act 3 of 2000 (PAJA) respectively.

There are also several new-generation environmental justice non-governmental organisations (NGOs) including the Federation for a Sustainable Environment (FSE) and Biowatch South Africa. The FSE, founded in 2008, promotes public participation and justice in environmental issues (the water case currently being litigated by FSE is discussed in part V). Biowatch is a research and advocacy group founded in 1997, which promotes biodiversity and sustainable livelihoods among – especially poor – rural communities. The litigation mounted by Biowatch South Africa is analysed in part IV as an example of how, despite the emerging holistic environmental justice movement, environmental litigation remains narrowly and exclusively cast. One of the most enduring and cross-cutting new-generation environmental justice organisations is SDCEA, which we have focused on because of its embedded history of socio-political struggle, industrial contestation, and high levels of environmental degradation and civic awareness.

III   CASE STUDY: SOUTH DURBAN COMMUNITY ENVIRONMENTAL ALLIANCE

The South Durban Community Environmental Alliance (SDCEA) is the leading community-based environmental organisation in Durban. Formed in 1995 out of apartheid struggles for social and economic justice, which were merged with opposition to large-scale industrial pollution in the area, SDCEA has been actively joining the dots between green, brown and red issues ever since. Across the South Durban industrial basin, SDCEA works to reveal the flaws in the Environmental Impact Assessment (EIA) processes, highlight blatant disregard for environmental laws by both industry and local authorities, and mobilise over other rights violations facing communities in the area, including labour-related issues and housing rights.[32]

(a)   Socio-economic and environmental context

The South Durban Basin (SDB) is approximately 24 kilometres long and four kilometres wide, and stretches in the north from Durban Harbour Point beyond the Umlazi Cuttings to the Isipingo Estuary in the south. It is home

---

32   C Barnett & D Scott 'Spaces of Opposition: Activism and Deliberation in Post-Apartheid Environmental Politics' (2009) 39 *Environment and Planning* 2612–31.

to approximately 300,000 people who live alongside various manufacturing and petro-chemical processing industries, and is known as the 'toxic hub of Africa'.[33] Like most urban areas in South Africa, the area comprises racially segregated suburbs – Umlazi and Lamontville are mainly African; Wentworth and Austerville are mainly coloured; Merebank and Isipingo are mainly Indian; and the Bluff is a largely white area.

While Umlazi, Lamontville, Wentworth and Austerville are predominantly poor and working-class suburbs, Merebank comprises more of a socio-economic mix. Housing approximately 10 per cent of Durban's Indian demographic, Merebank is a conglomeration of both lower and middle-class households, with spacious luxury homes occupied often by young professionals, alongside poverty-stricken households. Apart from the daily environmental challenges posed by their location in the SDB, issues affecting the poorer residents of the SBD include 'poverty and unemployment', and 'crime, drug abuse, and violence'.[34] Differing slightly, the Bluff, which is a blue-collar working-class white suburb, has lower un-employment. Yet, while it is not located directly adjacent to industry, prevailing winds carry soot and occasional oil spray from the refineries. Thus, notwithstanding the apartheid-inherited residential segregation, the high level of noxious pollution generated by industry[35] has created a platform for collective conscientisation, and the area has a long history of civic struggle and mobilisation.[36]

Industrial development was vigorously pursued in the SDB from the 1950s onwards and today the area is home to large oil refineries, paper mills and pulping plants and a sugar refinery. Allowed to expand virtually unchecked under apartheid, by the 1990s the area was acknowledged to have problematically high concentrations of air pollution.[37] According to a study completed in 2000, 11 stationary point sources (ie factories) were responsible for approximately 90 per cent of the total sulphur dioxide $(SO2)$[38] emissions for the SDB per year, with Sapref and Engen (the two major oil refineries), alone, contributing more than 65 per cent. These measures were based on

33 'South African Environmental Justice Struggles Against "Toxic" Petrochemical Industries in South Durban: The Engen Refinery Case' <http://www.umich.edu/~snre492/brian.html>.

34 See '2004 Crime Statistics for KwaZulu-Natal' <http://www.iss.org.za/CJM/stats0904/kzn. htm>.

35 Barnett & Scott (note 32 above) 2612– 31; R Nriagu, T Robins, L Gary, G Liggans, R Davila, K Supuwood, C Harvey, C Jinabhai & R Naidoo 'Prevalence of Asthma and Respiratory Symptoms in South-Central Durban, South Africa' (1999) 15 *European J of Epidemiology* 747–55; T Robins, G Batterman, U Lalloo, E Irusen, R Naidoo, B Kistnasamy, N Baijnath & G Mentz 'Air Contaminant Exposures, Acute Symptoms and Disease Aggravation Among Students and Teachers at the Settlers' School in South Durban' (2002) (interim report) <http://www.h-net. org/~esati/sdcea /positionpapers.html>.

36 D Scott & C Barnett 'Something in the Air: Civic Science and Contentious Environmental Politics in Post-Apartheid South Africa' (2009) 40 *Geoforum* 373–82.

37 SA Department of Environmental Affairs and Tourism 'South Africa Country Report' 14th Session of the United Nations Commission on Sustainable Development (2005) <http://www. environment.gov.za/sites/default/files/docs/sacountry_uncsdreport_atmosphere.pdf7>.

38 According to the US Environmental Protection Agency, sulphur dioxide (SO2) affects human health when inhaled, causes the most harm to the respiratory system and poses the greatest threat to people with pre-existing asthma or similar respiratory conditions, and can exacerbate asthma and lead to lung disease <http://www.epa.gov/airquality/sulfurdioxide/>.

the amount of reported daily fuel consumption reported by each company.[39] Today the skyline is littered with smokestacks and there is a pervasive smell of burning fuel in the air.[40] Toxic emissions into the air and water are a daily and cumulative threat to the residents and workers around South Durban, with air pollution creating the most proximate and pervasive harm, resulting in high rates of respiratory illnesses and cancers affecting the many residents of the affected communities.[41]

International studies have shown that childhood asthma disproportionately affects children living in poverty and residing in urban centres, many of whom are non-white ethnic groups. This is often attributed to the location of non-white residential areas close to high traffic densities and industry, which causes increased potential for decreased respiratory health caused by fumes from carcinogenic vehicle exhausts. Epidemiological studies conducted elsewhere around the world have correlated the presence of high traffic density with increased asthma symptoms, asthma hospitalisations, and decreased lung function.[42]

Specifically within South Durban, a preliminary study was undertaken at the Settler's Primary School located in Merebank in 2001.[43] This study was conducted in order to corroborate the linkages between air pollution and breathing problems amongst the students and teachers at the school, and was the first study to suggest the linkages between exposure to certain air pollutants emitted from the nearby refineries and mills, and the exacerbated poor respiratory status among children who have asthma. This study concluded that, although SO_2 levels were low compared with international standards[44] at the time of the study, there were a high number of students in grades 3 and 6 with asthma (52 per cent) of some kind, 26 per cent with persistent asthma (meaning asthma that causes symptoms more than about two times per week), and 11 per cent had asthma in the moderate to severe category. These figures are considered high compared with the asthma rate

---

39   E Cairncross 'Emission Inventory for South Durban' (2000) *Ecoserv Ref: DSEII* <http://www.h-net.org/~esati/sdcea/emissionscairncross.rtf>.

40   J Vidal 'Why South Durban Stinks of Rotten Cabbage, Eggs and Cat Wee' *Guardian* (6   December   2011)   <https://www.guardian.co.uk/environment/2011/dec/06/souht-durban-industrial-pollution>.

41   T Robins, G Batterman, G Mentz, B Kistnasamy, C Jack, E Irusen, U Lalloo, R Naidoo, B Kistnasamy, N Baijnath & H Amsterdam 'Respiratory Health and Air Pollution in South Durban: the Settlers School Study' (2005) 16(5) *Epidemiology* 79–104 <http://doeh.ukzn.ac.za/SDHealthStudy1299.aspx>.

42   For further reading see T Lewis, T Robins, J Dvonch, G Keeler, F Yip & G Mentz 'Air Pollution–Associated Changes in Lung Function Among Asthmatic Children in Detroit' (2005) 113 *Environmental Health Perspective* 1068–75 <http://www.ncbi.nlm.nih.gov/pmc/articles/PMC1280351/>; CA Aligne, P Auinger, R Byrd & M Weitzman 'Risk Factors for Paediatric Asthma – Contributions of Poverty, Race, and Urban Residence' (2000) 162 *American J of Respiratory Critical Care and Respiratory Medicine* 873–7 <http://ajrccm.atsjournals.org/content/162/3/873.full>.

43   Robins et al (note 41 above).

44   The average level of SO_2 in the outdoor air during the study period was 12 parts per billion (ppb). The average level of SO_2 over the entire previous year was 18.5 ppb. The World Health Organization (WHO) recommends keeping the average yearly levels at 19 ppb or less <http://www.who.int/mediacentre/factsheets/fs313/en/index.html>.

among South African children more generally.[45] As understood by SDCEA, the environmental problems caused by industrial pollution cannot be divorced from broader issues of environmental justice, including access to housing, water and health-care services, as well, ultimately, as South Africa's carbon-intensive growth-path.

(b) SDCEA's approach

Like most other geographic areas of environmental justice contestation in South Africa (such as in the Vaal triangle), opposition to the increase of industrial pollution combined with growing frustrations over unemployment, poverty and other socio-economic factors to create a social justice movement. Most of SDCEA's founding members have a history of community activism in the anti-apartheid movement and also the post-1994 housing movement. SDCEA first began in 1995 as a coalition of civic and resident associations from several communities who came together to push for stronger environmental regulatory standards in respect of the industries in South Durban. Gaining notoriety in the region over the years for being able to organise across the historic racial divisions of its communities including the Indian area of Isipingo and the conservative white area of the Bluff,[46] SDCEA has been able to speak out for environmental justice at local, national, and international levels with a real strong goal of justice for the poor. In recent years, SDCEA has increased its reach into some of the predominantly poor African areas such as Umlazi, which are often resistant to more traditional environmental outreach.

Although not a mass movement per se, SDCEA considers itself to be an important part of the larger environmental justice struggle within South Africa, as well as a key player vis-à-vis the global threat of climate change. SDCEA employs a wide range of tactics, but focuses on advocacy and has mounted sustained campaigns to force industry to adhere to environmental standards. One of SDCEA's successful strategies has been to build up detailed scientific expertise regarding pollution, and particularly the ability to monitor pollution levels through collecting and analysing samples,[47] and such knowledge has been shared within the communities to empower local residents and to increase SDCEA's legitimacy in the eyes of the state.[48] For example, in 2000, SDCEA's campaign against the air pollution produced by the Engen oil refinery, which involved working closely with residents and the local hospital and schools in the area to collect data on the number of illnesses

---

45   Robins et al (note 41 above).
46   Barnett & Scott (note 32 above).
47   Scott & Barnett (note 36 above).
48   Ibid 377. This began for SDCEA with Danish funding to undertake two initial scientific studies in the area, which gave statistics and figures to the harm that individuals were already aware of. These initial studies then flowed into 'the bucket brigade' sampling method, which was used as a means for community policing, wherein people living around the industrial complexes were for the first time, able to access accurate scientific measurements of chemical levels in the air they breathe through sporadic sampling of the air. These samples revealed the high levels of benzene, sulphur dioxide (SO2), and other harmful chemicals in the air, which helped to explain the high rates of Leukaemia and asthma in the SDB.

recorded, resulted in Engen introducing a technology that reduced its sulphur dioxide emissions by a quarter.[49]

SDCEA has also been successful at ensuring that the media takes up local issues and that there is constant pressure on industry and the government to promote environmental justice. For example, starting in 2000, SDCEA developed a 'bucket brigade' campaign, borrowing from the US the idea to get residents to collect and test samples of air pollution, and to expose pollution levels in the media. This tactic was seen both as a means of empowering local civilians by providing a simple tool for monitoring their own environment, and a powerful mechanism to attract journalists and generate a media storm. Through the bucket brigade, SDCEA has ensured that environmental issues are taken up by the media and brought to the attention of power-holders. In particular, this campaign has been successful at creating doubt in the public eye in terms of government and corporation's ability to adequately monitor industrial pollution.[50]

Alongside such tactics, SDCEA has also used litigation. Clearly, in its understanding and ethos, SDCEA pursues a combined approach to green, brown and red environmental issues. However, as we outline below, this approach has not yet been fully translated into SDCEA's litigation, which, to date, has remained quite narrowly cast.

(c)   SDCEA and the courts

Over the years, SDCEA has engaged litigation as a tactic and has mounted several cases in the courts. Somewhat surprisingly, this has overwhelmingly been in the form of narrow defences of traditional environmental protection, and largely through the lens of administrative law principles in which the environmental context is 'peripheral'.[51] One such case concerned the 2002 authorisation by the KwaZulu-Natal Department of Agriculture and Environmental Affairs for Mondi to construct and operate a 90-ton incinerator at its paper mill in Merebank, by virtue of granting the company that sold Mondi the relevant technology (Biotrace (Pty) Ltd) an exemption, in terms of s 28(A) of the Environment Conservation Act 73 of 1989, from various EIA criteria.

Having previously halted Mondi from continuing to dispose of excess industrial waste by embarking on concerted public advocacy campaigns that resulted in the closure of harmful toxic landfills in both Merebank and Umlazi,[52] SDCEA immediately sought an administrative review of this decision on the

---

49   H Corr 'CAMPAIGN HERO: Bongani Mthembu of South Durban Community Environmental Alliance' *Ecologist* (26 October 2011) <http://www.theecologist.org/how_to_make_a_difference/ cleaner_air_water_land/1106158/campaign_hero_   bonani   _mthembu_of_south_durban_ community_environmental_alliance.html>.

50   C Barnett 'Media Transformation and New Practices of Citizenship: the Example of Environmental Activism in South Durban' (2003) 51 *Transformation* 1–24 <http://www.history. ukzn.ac.za/ojs/index.php/transformation/article/viewFile/ 869/684>.

51   M Kidd 'Greening the Judiciary' (2006) 9(3) *Potchefstroom Electronic LJ (PELJ)* 72, 80.

52   N Schils 'South Africans Protest Mondi Paper's Multi-fuel Boiler, 2001–2006' (2011) Global Nonviolent Action Database <http://nvdatabase.swarthmore.edu/content/south -africans-protest-mondi-papers-multi-fuel-boiler-2001-2006>.

basis that the exemption was flawed and the construction would compound pollution in the community.[53] When the decision to construct the proposed incinerator was communicated, SDCEA requested Mondi to undertake a detailed study to determine the potential effects on air quality. This request was never obliged,[54] and SDCEA thus turned to the courts. In a press statement released on 18 September 2002, SDCEA announced that, with the assistance of the Legal Resources Centre (LRC) in Durban, it was pursuing 'legal action against government and Mondi' to 'demand an immediate halt to this attack on the poor...'.[55] On 6 May 2003, the Durban High Court found in SDCEA's favour, granting an interim interdict to halt Mondi's multi-million rand boiler pending the outcome of a judicial review.[56] On 9 July 2003, the Durban High Court ruled the exemption to be so 'palpably wrong' that it confirmed the finding of the court of 6 May 2003 to review and set aside the decision.[57] The judgment was seen by SDCEA as vindication of its broader campaign against Mondi's pollution. As Desmond D'sa, SDCEA executive director, reflects: 'we were very happy with the judgement'.[58]

Notwithstanding the immediate judicial victory, the Mondi case highlights some of the limits of pursuing environmental challenges through narrow, administrative law arguments. Indeed, the judgment has been criticised by Michael Kidd: 'the fact that the impugned decision concerned an environmental authorisation was incidental',[59] and the case can further be critiqued in that it did not delve into the broader socio-economic issues that are the core of SDCEA's environmental justice agenda. One of the critical consequences is that, by focusing on the defectiveness of the decision from a formal compliance/procedural perspective, Mondi was subsequently able to (relatively easily) rectify the non-compliance and proceed to obtain the necessary authorisation.[60] Mondi duly completed its multi-fuel boiler and its expansion project at its South Durban complex in 2006 and this is still in operation.[61]

More recently, SDCEA embarked on a campaign against eThekwini[62] municipality's proposed plan to build a new port extension. It is still early days, but litigation might be one of the tactics employed. If so, it will be interesting to see if a narrow environmental/administrative justice challenge is brought

53   'SDCEA Winds Mondi Combustor Interdict' (July 2003) 2 *SDCEA Community News* 1–2 <http://www.docstoc.com/docs/ 71434676/SDCEA-WINS-MONDI-COMBUSTER-INTERDICT>.

54   Schils (note 52 above).

55   'South Durban Communities Slam Government's Decision to Allow Mondi to Burn Coal' *South Durban Community Environmental Alliance* press statement (18 September 2002) <http://www.h-net.org/~esati/sdcea/pr18sept02.html>.

56   See, for example, Schils (note 52 above).

57   *South Durban Community Environmental Alliance v Head of Department: Department of Agriculture & Environmental Affairs KZN* 2003 (6) SA 631 (D).

58   Telephone interview between Anna Alcaro and Desmond D'sa, executive director of SDCEA (18 July 2012).

59   Kidd (note 51 above).

60   T Carnie 'Mondi Burner Gets Green Light' *The Mercury* (6 December 2004) <http://www.iol.co.za/news/south-africa/mondi-burner-gets-green-light-1.228731>.

61   Schils (note 52 above).

62   In 2002, the city of Durban began to be legally recognised by its Zulu name, eThekwini meaning 'the place of the bay'.

or whether SDCEA begins to try to broaden the legal claims to include, for example, issues around the likely threat to the livelihoods and socio-economic rights of the Airport Farmer's Association, including the likely loss of jobs to some 100-plus poor farm workers.[63]

If, as set out above, one of the main organisational promoters of a holistic environmental justice approach is not advancing such integrated environmental and socio-economic rights claims in its litigation, what does the rest of the environmental case law record reveal about cases that could potentially raise more integrated challenges? A related question is whether, especially if applicants are able to win cases through a narrow legal approach, it matters that litigation has not engaged or reflected a more holistic understanding of environmental rights. Below we analyse some of the key cases brought by environmental movements in order to reflect on these questions. Our coverage is not exhaustive, but the cases we highlight illustrate some of the characteristics and contours of contemporary environmental rights litigation in South Africa.

IV ENVIRONMENTAL CASE LAW

Although vindication of rights does not necessarily entail litigation and courts, litigation offers a potentially powerful platform for not only resolving disputes but also highlighting persistent problems. Moreover, in a constitutional democracy such as South Africa's, courts can play an important role in interpreting and enforcing constitutional rights:

> As such, it falls on the judiciary to use its powers of judicial review to assess the act and conduct of the legislature and executive for consistency with the Constitution. In a Constitution such as the South African Constitution, which provides for the horizontal application of human rights, it is the role of the judiciary to enforce the right against natural or juristic persons. It is through this process of judicial review that courts can provide content to the right and address problems such as vagueness and lack of definition... It is therefore only fit that the judiciary should similarly develop environmental rights.[64]

However, as noted by Feris, 'when one examines South African jurisprudence, there seems to be a marked dearth of cases where the environmental right has been fully utilised and clearly interpreted'.[65] There may be several reasons for this gap, each which might go some way to explain why environmental justice movements such as those mentioned above have thus far pursued such narrow environmental claims.[66] First, at the level of the courts, there may be a discomfort with the subject matter or of adjudicating rights in an integrated way. Second, at the level of the lawyers, there may be a deliberate narrowing of claims to traditional, 'winnable' points of law. Third, at the level of litigants – and beyond the knock-on effect of the first two factors, ie

---

63 R Umar 'Dug-out Port SpellsDdoom for Farmers' *Daily News* (27 April 2012) <http://www.iol. co.za/dailynews/news/dig-out-port-spells-doom-for-farmers-1.1285390#.UCq0I5ibK5c>.

64 Feris (note 4 above) 37–8.

65 Ibid 38.

66 We leave out of these possible reasons the argument that law and courts are inherently unable to deal with complex structural issues. While this position can certainly be argued, we take the view in this chapter that litigation is the art of the possible. As such, we are interested in an empirical examination of actual litigation to attempt to pinpoint some of the obstacles and missed opportunities in developing an inclusive environmental justice jurisprudence.

in deciding whether and what to litigate, litigants are likely to be influenced by lawyers who, in turn, are influenced by the approach of courts – there may be the issue that environmental justice litigation often creates perverse incentives and contradictions for different classes of litigants especially where poor people have to choose between livelihoods and pollution.

Regarding the first possible explanation, while it is not possible to delve into this consideration without intensive interviews with judges – and notwithstanding some evidence from socio-economic rights cases themselves that judges are uncomfortable with genuinely integrated approaches to rights[67] – there are some indications from recent judgments (albeit in relation to economic development rather than socio-economic rights per se) that at least some judges might be willing to pursue an inter-connected approach to environmental rights. For example, in *Fuel Retailers*,[68] the Constitutional Court stressed:

> The Constitution recognises the interrelationship between the environment and development; indeed it recognises the need for the protection of the environment while at the same time it recognises the need for social and economic development. It contemplates the integration of environmental protection and socio-economic development. It envisages that environmental considerations will be balanced with socio-economic considerations through the ideal of sustainable development. This is apparent from section 24*(b)*(iii) which provides that the environment will be protected by securing 'ecologically sustainable development and use of natural resources while promoting justifiable economic and social development'. Sustainable development and sustainable use and exploitation of natural resources are at the core of the protection of the environment.[69]

However, despite the Constitutional Court's clarification in *Fuel Retailers* of the integration within sustainable development of the demands of economic development, social development and environmental protection, the vexed question remains of how we should 'interpret sustainable development in a country which faces large scale poverty and where such a clear and unequivocal need for economic and social development is present'.[70] Yet it is critical that the courts do begin to address this interpretational issue. This is because the issue of intra-generational equity flagged in s 24(b) of the Constitution, along with the distributional demands of socio-economic justice especially in as unequal a country as South Africa, underpins environmental degradation and demands an answer within the environmental rights paradigm.[71]

On the second explanation, it is trite that, outside of the more activist courts found in, for example, Latin America and India, courts usually are unwilling to make more of a case than is presented to them on the papers. This means that judges rely to a great extent on the way cases are framed and claimed by lawyers, who are traditionally wary of taking risks especially when there is a narrow, winnable point to be taken. While further research is

---

67   It is a curious characteristic of South African socio-economic rights judgments especially at the Constitutional Court level that, while many seem to promote an integrated approach as part of the background descriptions, none has pursued an integrated approach in its remedies and orders.

68   *Fuel Retailers Association of Southern Africa v Director General Environmental Management, Department of Agriculture, Conservation and Environment, Mpumalanga Province* 2007 (6) SA 4 (CC).

69   Ibid para 45.

70   Feris (note 4 above) 41.

71   Ibid.

needed to delve into the choices lawyers make, and especially to investigate whether any compromises were made from the perspective of the instructing environmental movements, from a facial examination, the case law reflects a remarkably narrow range of issues focusing on the PAIA and the PAJA-related applications or other forms of administrative law challenges. As mentioned above, VEJA's planned litigation is focused on a series of PAIA applications, and one of the key legal victories for SDCEA, the case against the Mondi incinerator, was essentially an administrative law case. Similarly narrow administrative law approaches characterise two other critical cases brought by otherwise more inclusively cast environmental justice movements: the litigation brought by Earthlife Africa to challenge the government's project to develop nuclear energy via the Pebble Bed Modular Reactor (PBMR); and the litigation brought by Biowatch to obtain information about the pursuit of Genetically Modified (GM) crops. In the *Earthlife Africa* case,[72] Earthlife Africa brought a judicial review (under s 36 of the Environmental Conservation Act 73 of 1989,[73] read with s 6 of PAJA[74]) that successfully set aside a highly controversial decision by the Department of Environmental Affairs and Tourism and Eskom to develop a PBMR project at the site of the Koeberg nuclear power station near Cape Town on the basis that the EIA process did not involve the necessary consultation with interested parties. Although the litigation was instrumental in scuppering the PBMR plans, it was an extremely narrowly constructed case. As explained by Kidd, the crux of the matter was that there was a failure of audi alteram partem and the environmental nature of the administrative decision was not a critical aspect of the case or the decision.[75] Nor did the legal team raise any of the other broader impacting environmental issues that Earthlife Africa espouses such as the natural systems and health-related concerns over nuclear power, the need for South Africa to invest in renewable energy etc.

The *Biowatch* case[76] was curious in that what started out as a case about trying to hold the government accountable for the decisions it made in relation to GM crops (again, a PAIA request for the record of the decision) ended up as a case about the awarding of costs orders in public interest matters.[77] The litigation sequence prior to the ultimate judgment of the Constitutional Court is explained in Biowatch's February 2009 newsletter as follows:

> Biowatch is a South African non-profit organisation, that aims to ensure the just use of biodiversity and sustainable approaches to agriculture. Over the past eight years Biowatch has found itself in an extraordinary situation. What started in 2000 as a seemingly simple request for information from the Department of Agriculture about the status of genetically modified (GM) crops, has resulted in a

---

72   *Earthlife Africa (Cape Town) v Director-General: Department of Environmental Affairs and Tourism* 2005 (3) SA 156 (CC).
73   Environmental Conservation Act s 36 deals with judicial review.
74   PAJA s 6 of sets out the requirements for judicial review of administrative action.
75   Kidd (note 51 above).
76   *Trustees for the time being of the Biowatch Trust v Registrar; Genetic Resources* 2009 (6) SA 232 (CC).
77   For an analysis of the judgment see, for example, T-L Humby 'The *Biowatch* Case: Major Advance in South African Law of Costs and Access to Environmental Justice' (2010) 22 *J of Environmental Law* 125.

lengthy legal process with Biowatch fighting for its survival. Although Biowatch substantially won the case in the Pretoria High Court, and the right to access this information, it was ordered by the judge (in a strange twist of events) to pay the costs of the multi-national company Monsanto, which had joined forces with the Department of Agriculture against Biowatch. This decision was upheld on appeal, although not with unanimous judgment. Biowatch is now preparing a further appeal for the case to be heard in the highest court of our land, the Constitutional Court.[78]

So, a narrowly framed access to information request (initially lodged prior to PAIA so grounded in s 32 of the Constitution) became broader in the course of the litigation. Nonetheless – although the Constitutional Court ultimately ruled in favour of Biowatch and the case had the unanticipated, positive outcome of determining that costs will generally not be awarded against organisations that bring litigation in the public interest – the value of the litigation is not so clear in relation to the issues that actually motivated the case.[79] Moreover, notwithstanding the fact that the litigation opened up debates about GM crops, it is arguable that pursuing a narrow access to information case became a sideshow, leaving the underlying issues largely unaddressed. While highlighting that the Biowatch case catalysed a new era of openness in relation to information about GM crops, Rachel Wynberg and David Fig point to the limits of the litigation:

> It is debatable as to whether or not the information that was obtained facilitated environmental protection. Despite obtaining access to a substantial amount of information, the practical use and relevance of this information was limited. Most notably, by the time information was received by Biowatch it was six years after the initial request. During this period more than 130 permits had been approved, a further 108 applications received, and more than one million hectares planted to GM crops in South Africa. At the same time, all records relating to GM crops between 1991 and 1999 had been disposed of, precluding any access to all this information.[80]

Turning to the third explanation, notwithstanding any apparent synergy between environmental and socio-economic issues, there are clearly serious tensions raised in the course of litigation. This is especially the case where the issue concerns pollution and/or health-care related damage to surrounding communities. Where communities are poor, the temptation to settle for even paltry amounts of money or simply to retain jobs is understandably high. Writing about the struggles of the Steel Valley Crisis Committee (SVCC), which emerged in 2002 to try to interdict ISCOR from polluting the groundwater and thereby resulting in loss of livelihoods and livestock as well as serious health problems, Cock explains that the action to interdict ISCOR initially appeared to illustrate the capacity of environmental issues to overcome the racial and class divisions between victims (of black and white race groups), and 'unite their "particularistic identities" in a common cause'.[81] However, ISCOR was able to divide and buy off the majority of the litigants and the litigation disintegrated. Cock, together with Victor Munnik, document how – in the wake of a successful case brought by a local

---

78 Biowatch South Africa 'Biowatch at the Constitutional Court' *Biowatch Briefing* (February 2009).

79 This critique should not be taken to suggest that we regard the *Biowatch* litigation, or any of the cases discussed in this chapter, as having been misplaced. Certainly most, if not all, the cases have had valuable impacts. We are more interested here in assessing whether the litigation could have gone even further and had more ambitious results.

80 Wynberg & Fig (note 5 above) 18–9.

81 Cock (note 12 above) 1.

resident, Johnny Horne, as leader of an all-white group, in which ISCOR paid out millions of rands in damages in an out-of-court settlement – a multi-racial group of 16 applicants, all owners of smallholdings, emerged to attempt to bring further litigation against ISCOR. This time, 'ISCOR reacted differently' – it disowned its own technical consultants' reports, failed to admit liability and imposed a gagging order on the 16 litigants'.[82] Instead, ISCOR offered monetary compensation to the applicants and all but two accepted the compensation, not least as ISCOR argued that, if the interdicts were granted, it would have to close operations, thereby rendering thousands of people unemployed. In the words of one of the litigants, 'we want ISCOR to clean up without closing down. We don't want to see our brothers and sisters out of jobs'.[83] Cock and Munnik explain how the issue of compensation divided the previously united group:

There were rifts in the community, as to the desired outcome of the court case. Half the community wanted ISCOR to pay them out for their properties so that they could move to a new area while the other half wants to stay in the area and wants ISCOR to upgrade its plants and piping in order to prevent further ground water contamination.[84]

Analysing the litigation, the authors come to the conclusion that in contrast to the initial hopes that the case would unite the various communities under a common threat, in fact it failed to penetrate deeply enough into any of the communities to withstand the pressure of the offers of compensation.[85] According to a candid appraisal of a member of the legal team:

The community got divided when the judge dismissed the case. Some settled and some didn't. People were very angry. The lawyers were fighting among themselves. The lawyers gave people different advice on whether to accept the settlement or not...[86]

Defeats such as those experienced by the SVCC highlight the difficulties and contradictions in attempting to mount cross-class environmental justice litigation. In this instance, ISCOR bought out desperate litigants who, ultimately needed employment and money more than they needed a sustainable environment. Similarly, in discussing the SDCEA Mondi case, Sharad Chari details several interesting points, of which he discussed how 'the possibility of a negotiated settlement exposed contradictions between legal counsel and community activists, as the latter would not give up its militancy even for important concessions from the industrial giants'.[87]

---

82   J Cock & V Munnik 'Throwing Stones at a Giant: an Account of the Steel Valley Struggle Against Pollution from the Vanderbijlpark Steel Works' *Centre for Civil Society (CCS) Research Report* (June 2006) 17–8.
83   Ibid 18.
84   Ibid 22.
85   Ibid.
86   Ibid.
87   S Chari 'Political Work: The Holy Spirit and the Labours of Activism in the Shadows of Durban's Refineries' (2004) *Centre for Civil Society (CCS) Research Report* 30 <http://ccs.ukzn.ac.za/files/RReport_30.pdf>.

V   CONCLUSION: TOWARDS AN ENVIRONMENTAL JUSTICE JURISPRUDENCE?

From our examination it does seem that environmental organisations have been playing it safe; going for the winnable points in court and not really pushing the boundaries of s 24 of the Constitution, let alone venturing into the brown and red components contained in the socio-economic rights clauses. Moreover, as our analysis highlights, even where narrowly framed litigation results in judicial wins, such victories might be pyrrhic, leaving the underlying problems unaddressed, as was the case in relation to SDCEA's litigation against Mondi and Biowatch's litigation against the government and Monsanto. From the perspective of litigation brought by environmental organisations and NGOs, it appears that there is still a wide divide between green, brown and red issues. It should be noted that socio-economic rights organisations, too, are not building the legal bridges towards an alliance with environmental rights. For example, the only water rights case to have come before the Constitutional Court, *Mazibuko*,[88] did not include environmental rights arguments even though it involved the 'environmental' issues of the domestic use of water by multi-dwelling poor households with waterborne sanitation.

However, it is not unlikely that, as environmental justice movements mature and/or the litigating organisations mature, more complex cases might be brought that attempt to expand the grounds for and scope of environmental litigation. For example, in 2012 two cases were launched by FSE involving a cross-over between environmental and water-related rights. While not yet concluded, these two cases – *Mjadu v The National Nuclear Regulator;*[89] and *The Federation for Sustainable Environment v The Minister of Water Affairs*[90] – both involve complex and integrated arguments about the impact of mining operations (specifically Acid Mine Drainage) on local environments for both human and natural systems, and both involve an environmental justice NGO (Federation for a Sustainable Environment), along with socio-economic rights' litigation NGOs (LRC and SERI). Although too early to assess, it is possible that these cases might signal a new approach to environmental cases that seeks to combine environmental and socio-economic rights and infuse an expansive and holistic approach to environmental rights litigation. In the context of environmental and climate change and its inevitable pressure on resources, environmental and social justice issues are increasingly indivisible[91] and such a development is critical. As such, environmental and socio-economic rights organisations, lawyers and judges all have a responsibility to contribute towards developing a more integrated and progressive interpretation of environmental law.

---

88   *Mazibuko* (note 3 above).
89   Unreported case number 24611/2012 (South Gauteng High Court).
90   Unreported case number 35672/12 (North Gauteng High Court).
91   Cock (note 12 above) 30.

RED-GREEN LAWFARE?
CLIMATE CHANGE NARRATIVES
IN COURTROOMS

CATALINA VALLEJO AND SIRI GLOPPEN

I   INTRODUCTION

> [T]he worlds legal systems, both international and national, have never seen a challenge like climate change. The science involves complexities of global ecology that are of a scale new to the courts[1]

In this chapter we analyse how environmental and social problems connected to climate change are being raised in court cases in various jurisdictions.[2] Our aim is to better understand the constraints and possibilities of climate change litigation and to see whether and how economic, social and cultural (ESC) rights feature in the legal arguments and in the design of remedies. Particular focus is on ecology and equity problems associated with energy production from fossil fuels, which is the major cause of anthropogenic greenhouse gas emissions (GHGs).

We use the notion of 'narratives' to explore and make sense of climate change cases brought to court in different countries.[3] The cases vary in terms of the actors involved, the claims and judicial responses and contain valuable insights about the complex web of socio-legal problems arising from a changing climate. Our aim is to retrieve some of the discourses that conceptualize political issues as legal storylines,[4] and we use the narrative perspective to help our understanding of how various issues form part of the climate change story of different groups going to court. For the purpose of this chapter, we use it concretely to determine whether and how ESC rights and environmental law are being integrated to form what we call *red-green lawfare*.

The first part of the chapter introduces the concept of *red-green lawfare*, and places climate change litigation in context, it outlines relationships

---

1   DB Hunter 'The Implications of Climate Change Litigation for International Environmental Law-Making' (2007) 2.
2   For an overview and typology of cases, see also S Gloppen & AL St. Clair 'Climate Change Lawfare' (2012) 79 *Social Research* 899.
3   We use narratives in the sense of 'stories' that have plots and different actors, and that are embedded in complex socio-political contexts. Thus we build on and expand the linguistic analysis of narratives in K Fløttum & Ø Gjerstad 'The role of social justice and poverty in South Africa's National Climate Change Response White Paper' (2013), in this volume, according to which plots recount a problem or complication, followed by a sequence of events or actions, which take place to achieve particular effect(s). See also K Fløttum & Ø Gjerstad 'Arguing for Climate Policy through the Linguistic Construction of Narratives and Voices: The Case of the South African Green Paper "National Climate Change Response"' *Climatic Change* (2012) (3-4) 115.
4   See for instance ibid.

between climate change, energy production and socio-economic development and highlights harms relevant to the law that this can implicate.

The second part of the chapter examines the growth of climate court cases since the early 2000s, in a context of failure to provide a regulatory framework for effective and fair mitigation and adaptation, both in international law and domestic legal systems. We discuss a selection of cases brought to court in various jurisdictions, identifying actors, arguments and courts' responses. The cases fall in three main categories: *civil law (tort)* cases, seeking compensation from fossil-fuel corporations for climate-related damages; *administrative law* cases, seeking regulation from State agencies, typically of GHG emissions and standards for environmental impact assessment; and *public international law* cases, demanding protection for communities most vulnerable to climate related harms. Towards the end of the chapter we discuss some of the emerging strategies in climate litigation that advocates are discussing as possible ways forward.

II   RED-GREEN LAWFARE AND THE CONTEXT OF CLIMATE CHANGE LITIGATION

Climate change litigation is located at the intersection between environmental (green) and socio-economic (red) issues, bringing to the forefront complex socio-ecological relationships.[5] It involves elements of environmental law (claims for the reduction of global warming and the protection of biodiversity), but also comprises elements of social justice. If populations stricken by poverty and social marginalization are the most vulnerable to the impacts of climate change, strategies to address the climate problem must also seek adaptation and transformation possibilities for people lacking the material basis for a life with dignity.[6]

As pointed out by Dugard and Alcaro in this volume, there is tension between environmental and social rights litigation.[7] At times 'environmentalism has operated effectively as a conservation strategy that neglected social needs,'[8] typically involving the protection of animals, plants and other natural resources. Other forms of environmental litigation have been concerned with the social effects of environmental harm, such as chemical contamination, toxic wastes, and small particles. Interestingly, in the United States (USA), such cases largely affected black low-income communities. According to Foreman, once this connection was made, the rhetoric evolved from mere conservationist concerns to 'environmental equity,' which morphed into the more provocative 'environmental racism' and eventually 'environmental justice,' which sought to fuse class and race-based complaints.[9]

---

5   On socio-ecological relationships in the context of climate change see for instance S Sullivan *Green: Going Beyond 'the Money Shot'* (2011) <http://siansullivan.net/>.

6   C Foreman 'On Justice Movements: Why They Fail the Environment and the Poor' (2013) Winter *Breakthrough Journal*.

7   J Dugard & A Alcaro 'Lets work together: Environmental and socio-economic rights in the courts' (2013) in this volume.

8   J Cock *Connecting the Red, Brown and Green: The Environmental Justice Movement in South Africa* (2004) 5. Cited in Dugard & Alcaro (note 7 above).

9   Foreman (note 6 above).

Cock argues that the notion of environmental justice represents an important shift away from the traditional concept of environmentalism, essentially concerned with the conservation of threatened wilderness, to include socio-economic or 'red' issues. In this sense, Cock understands the environmental justice movement as 'bridging ecological and social justice issues' in that it puts the needs and rights of marginalized communities at the centre of its concerns.[10] Environmental justice then is located at the confluence of various challenges: the struggle against poverty and inequality and the struggle to protect the environment, per se and as the natural resource base on which all economic activity depends.[11] Drawing on this, we frame the courtrooms narratives presented here within the concept of climate justice.

But before analysing liability and remedies sought through litigation, let us revisit the problem the court cases seek to address, that is, the problem of climate change and its causes, 'which is mainly focused around the energy resource question.'[12]

Energy can be seen as the primary basis of an economy and societies give priority to reliable, low cost energy production in which electricity, gasoline, coal or hydrogen are converted into a vast range of devices and processes.[13] The extraction and burning of fossil fuels (oil, gas, and coal) on which most economies depend to produce energy, create greenhouse gases (GHGs). Forests and oceans can to some extent absorb these, but excess gases accumulate in the atmosphere causing a greenhouse effect resulting in global warming and other ecological changes such as ocean acidification.[14] Energy-related carbon dioxide (CO_2) represents the greater part of anthropogenic GHG emissions.

Climate research indicates with increasing certainty that a warmer planet causes multiple alterations in weather patterns, such as colder winters and warmer summers, extreme patterns of flood and drought, sea level rise (affecting coastal land and small islands), changes in precipitation, early melting of ice packs, erosion, and higher frequency and force of storms amongst other phenomena. This set of weather alterations is what is known as climate change.[15]

While all humans are susceptible to a range of weather events depending on the place they inhabit, possibilities for resilience and prevention of large damage vary according to available infrastructure and resources. It is increasingly clear that the adverse socio-economic effects from climate

---

10   Cock (note 8 above).
11   At present, the human population and the average energy consumption are increasing, while the total area of productive land and natural resources are fixed or declining. WE Rees 'Revisiting Carrying Capacity: Area-Based Indicators of Sustainability' in AR Chapman, RL Petersen & B Smith-Moran (eds) *Consumption, Population, and Sustainability: Perspectives from Science and Religion* (2000) 72.
12   J Marburger 'A Global Framework: International Aspects of Climate Change' (2008) 30 *Harvard Int Rev* 48.
13   Ibid 48-51.
14   See Intergovernmental Panel on Climate Change (IPCC) Fourth Assessment Report: Climate Change 2007 (AR4); IPCC Managing the Risks of Extreme Events and Disasters to Advance Climate Change Adaptation 2012 (SREX).
15   Ibid.

change are likely to be very substantial and affect crucial sectors such as water resources, agriculture, forestry, fisheries, health and human settlements, with low-income countries being the most vulnerable.[16] Middle and low income countries with high levels of inequality have lesser capacity to adapt to significant weather alterations, and the poorest people are most vulnerable to climate change related harms, as they are to other strains.[17]

Aside from uneven resilience and possibilities for adaptation to climate events, climate change deepens other equity problems. Since the industrial revolution fossil fuels have been considered the cheapest source of energy for large-scale economic activity and their use continues to grow at an exceptional speed.[18] But this carbon-led growth is both unsustainable and unequal. Positive gains of fossil energy production – for instance in terms of time-efficient modes of transport, appliances and other comfort devices in everyday life – are disproportionally benefiting high income populations, while socially and/or economically marginalized populations suffer disproportionally from the negative side effects of the fossil fuel industry – including effects other than those associated with climate change, such as oil spills, gas flaring, air pollution, extractive-related displacement, deforestation, and restrictions on peoples' self-determination. Development processes are accelerating also in the global south, but billions are left behind. More than one billion people still lack access to electricity due to costs.[19] Re-negotiating the terms of industrialized economies is thus important both from the perspective of the environment and for reasons of socio-economic equity.[20]

The question of climate change is thus not only a matter of nature conservation, for example through new energy technologies, but also a question of socio-economic equality in the production and enjoyment of those technologies. Studies suggest that high-income societies consume on average about three times their fair share of sustainable global output.[21] At the same time, not only high income, industrialized countries have high levels of GHG emissions and energy consumption. As developing countries like China, India, Brazil and Malaysia industrialize; they also increase their GHG emissions,

16 Intergovernmental Panel on Climate Change *Presentation of Robert Watson, Chair IPCC, at the Sixth Conference of the Parties to the United Nations Framework Convention on Climate Change, The Hague* (13 November 2000). Cited in S Olmos *Vulnerability and Adaptation to Climate Change: Concepts, Issues, Assessment Methods* (2001); See also IPCC report AR4 2007 (note 14 above).

17 See IPCC reports AR4 2007 and SREX 2012 (note 14 above). See also K O'Brien, AL St. Clair & B Kristoffersen (eds) *Climate Change, Ethics and Human Security* (2010).

18 The central role of energy is also causing governments to intervene through regulation, taxation, or incentives pursuing greater efficiency in the use of energy. Marburger (note 12 above).

19 Foreman (note 6 above).

20 In this regard, Kumi Naidoo (South African member of the Anti-Apartheid movement and leader of Greenpeace International in 2009) has asserted that the struggles against global poverty and climate change may be two sides of the same coin. 'Look, 1.6 billion people have no access to energy and yet live in regions that are blessed with abundant solar, wind, wave, and geothermal energy (...). Traditional western-led environmentalism has failed to make the right connections between environmental, social, and economic justice.' Naidoo in J Broder 'Greenpeace Leader Visits Boardroom, without Forsaking Social Activism' *New York Times* (6 December 2011). Cited in Foreman (note 6 above).

21 See for instance, Rees (note 11 above).

while at the same time moving millions of people out of poverty.[22] Climate change thus lies at the core of energy and development policy-making.[23] From a legal perspective, technologies for energy production and usage not only need to be assessed from a conservationist or 'green' viewpoint, but also from socio-economic and cultural perspectives. In this chapter we will examine to what extent, and how litigation is dealing with this nexus, forming what we call red-green lawfare.

(a)   In search of solutions: the move to litigation

What are possible and feasible approaches to the climate related problems outlined above? The main goal of the international legal regime is to reduce GHG emissions (climate mitigation).[24] Current consensus suggests that to achieve this, at least three lines of response are required: changing, diversifying and/or adapting technologies for energy production; increasing the capacity of the biosphere to absorb CO_2;[25] and reducing levels of energy usage through lowering consumption and improving energy efficiency technologies. Currently emissions levels are not decreasing. And even if mitigation efforts should succeed, the effects are not immediate and there is urgent need to facilitate resilience or adaptation to climate events, including through transnational allocation of resources for assistance to affected communities.

While the challenges cry out for collective action to rectify the malfunctioning of our economic and social systems, current governance structures are not well equipped for dealing with global trans-boundary and intergenerational issues. As discussed above, carbon is deeply embedded in the global economy, with its positive and negative effects manifesting in specific ways in different places and at different times. This makes climate change at the same time an individual, local, state, national, regional, and international problem.[26] Regulatory difficulties at all levels show how intensely CO_2 and other GHGs are rooted in the ways of life of most societies.[27] Osofsky concludes that climate change has the character of a multiscalar social problem which cannot be approached effectively – whether through political or legal avenues – by the use of existing national governance frameworks only.[28]

Current international legal frameworks to address climate change are weak. The main international instruments are the 1992 United Nations Framework Convention on Climate Change (UNFCCC, in force since 1994), and the Convention's Kyoto Protocol (1997, in force since 2005). The UNFCCC

---

22   Foreman (note 6 above).
23   See for instance, International Energy Agency (IEA) 'Climate Change' (2013) <http://www.iea. org/topics/climatechange/>; World Bank 'Development in a Changing Climate: Making Our Future Sustainable' (2013) <http://blogs.worldbank.org/climatechange/>.
24   Ibid; J Lin 'Climate Change and the Courts' (2012) 32 *J Legal Stud* 35.
25   Marburger (note 12 above).
26   HM Osofsky 'Is Climate Change 'International'? Litigation's Diagonal Regulatory Role' (2009) 49 *Va J Int'l L* 585.
27   Ibid.
28   Ibid.

provides a framework in international law for multilateral cooperation to tackle climate change, including obligations of industrialized states to report on and reduce their CO_2 emissions, conduct adaptation measures, and provide assistance to non-industrialized countries enabling them to adapt to the adverse impacts of climate change.

Over 190 countries plus the European Union are Parties to the Kyoto Protocol. This includes all member states of the UN, except Afghanistan, Andorra, Canada, South Sudan and the USA. The USA signed Kyoto but did not ratify, and Canada withdrew in 2011. Australia only ratified the protocol in 2007.[29] By 2013, many refer to the UNFCCC as a failure, not making any difference to carbon emissions, which are still increasing.[30]

While the UNFCCC is the centre of attention in legal discourse on climate change, climate related petitions have also been filed in other forums articulating issues of justice and equity that have been relegated from the dominant discourses,[31] bringing climate change to the forefront in socio-legal debates.

Most climate change court cases are filed in the USA, Australia and the European Union (EU), very few in the global south.[32] In high emitting countries where the government has refused to ratify the Kyoto protocol or to take mitigation and adaptation steps, litigation has served as a tool to fight administrative blockage. Civil society groups facing obstructions to successfully work the political apparatus, or wanting to complement their lobbying campaigns have taken cases to court.[33] But, why don't we see a significant wave of litigation in countries of the global south, who are the ones facing more difficulties in terms of adaptation and resilience?

The following elements may contribute towards a tentative, initial answer: Firstly, as we will see, cases in the global north have been mainly towards national governments and the domestic industry. Global south countries have no Kyoto or other legal obligations to reduce emissions; making it harder to argue the breach of a legal obligation, and the energy industry in non-industrialized countries has had less impact on global warming, which makes liability claims difficult. The industry is also largely perceived and promoted as a necessary and legitimate means towards development in contexts of poverty and inequality. Secondly, there are legal difficulties involved in bringing suits against transnational corporations in domestic global south

---

29 During the Conference of Parties (COP) 13 in Bali, on 3 December 2007.

30 See for instance, B Amundsen & E Lie 'Why the Kyoto Agreement Failed' (2010) <http://www.forskningsradet.no/en/Newsarticle/Why_the_Kyoto_agreement_failed/1253963392536>; S Connor 'Scientists Say Kyoto Protocol Is "Outdated Failure"' *The Independent* (25 October 2007).

31 Lin (note 24 above).

32 In European courts we see a number of cases filed in the context of the EU Emissions Trading Scheme (ETS), seeking to make it effective by refining new market mechanisms. These cases present significant challenges for the courts, asked to decide on complex design regulation issues with economic as well as political implications for the EU's responses to climate change. NS Ghaleigh 'Emissions Trading before the European Court of Justice: Market Making in Luxembourg' in D Freestone & C Streck (eds) *Legal Aspects of Carbon Trading: Kyoto, Copenhagen and Beyond* (2009). Cited in Lin (note 24 above) 49.

33 Lin ibid. In the same sense see WCG Burns 'Introduction' in WCG Burns & HM Osofsky (eds) *Adjudicating Climate Change: State, National, and International Approaches* (2009).

courts for harms of a global nature (i.e. problems to demonstrate standing, jurisdiction, causality and violation of a duty of care). While some cases in the south address environmental and human rights harm by transnational energy companies at the local level, these, as we will see, use climate change merely as a supporting argument. Lastly, claims based on climate justice arguments, seeking reparations from other states, have so far not had success in international courts, and the lack of a global environmental court might have hampered the development of jurisprudence in this area.

Meanwhile in the USA, the resort to adjudication in the early 2000s should be seen against the refusal of the Bush Administration to support climate change regulations or policies, causing deep frustration among concerned people. In addition to the adverse political opportunity structure (symbolized by the government's refusal to ratify the Kyoto protocol), there was a favourable legal opportunity structure (including the presence of well-funded environmental groups, a tradition of public interest litigation, and an active civil society).[34] Plaintiffs used litigation in various ways seeking to push state regulation or hold corporations accountable for their contribution to global warming – but also simply 'to maintain the climate change issue on the political agenda.'[35]

II   Typology of Climate Related Lawfare

Lawsuits to address climate change fall into three general categories according to the narratives on responsibility put forth.[36]

The first category is **civil law** claims, or what we call *climate tort lawfare*. Cases have been filed claiming that corporations' GHG emissions have caused concrete harms that should be repaired. The dominant narrative in these cases is that corporations are damaging the environment by nuisance or negligence by producing large GHG emissions, and therefore should pay damages to affected groups, and reduce their emissions.

The second category is **administrative law** claims, or *climate regulations lawfare*, claiming that governments should take (or not take) certain actions to mitigate or adapt to climate change. The narrative here is that governments are damaging the climate (usually by omission), and should take affirmative measures to protect human rights, the sustainability of the environment and the rights of future generations.

The third category is **public international law** claims or *transboundary climate lawfare*. Petitions are filed before various international bodies regarding the adverse effects of climate change on indigenous peoples, vulnerable communities with poor adaptation capacity, and places considered world's heritage. The dominant narrative running though these cases is that governments and corporations most responsible for global emissions have an obligation to transform their energy policies and assist communities in other

---

34   Lin ibid.
35   Ibid 35.
36   We develop this trilinear typology from MB Gerrard *Global Climate Change and United States Law* (2007) 21.

countries suffering climate related harms of no fault of their own, and lacking own means to adapt.

Below we focus on what can be seen as paradigmatic climate change cases in each of these categories. The cases engage one or more of the following 'intractable' features of climate change: The *transnational and free-rider character* of the problem (causes as well as consequences of climate change cross all borders, those who suffer most have often contributed the least, no political authority 'owns' the problem and potential solutions give rise to collective action problems); the *long time horizon* (most consequences – both of current emissions and of costly mitigation measures – materialize much later, hence many current decision-makers might not suffer the harm of climate change, or experience the benefits of mitigation efforts, only bear the costs of the latter); *attribution problems* (while there is broad scientific agreement that the climate is changing, natural variability and scientific uncertainty make it difficult to determine effectiveness of mitigation and adaption measures – and even more so, to attribute responsibility for particular harms).

(a) Climate tort lawfare: suing industry for damages through civil claims

In various cases, particularly in the USA, litigants have sued coal, oil, gas and other large industrial corporations responsible for large GHG emissions, asking the court to declare that their economic activity has damaging effects on the environment and human communities. The central question in these cases is 'whether companies can cause damages through greenhouse gas emissions and if they can be held accountable.'[37]

In *Connecticut et al v American Electric Power Co et al* (131 S Ct 2527 [2011]), the City of New York, and three advocacy groups went to court against five electric power producers under the federal and state common law of nuisance. The plaintiffs alleged that these companies, as the five largest emitters of CO_2 in the USA were responsible for substantial CO_2 emissions and therefore liable for public health and environmental harms stemming from global climate change. The plaintiffs asked the court to order limits and phased reductions of the defendants' CO_2 emissions. The court rejected the claims as *non-judiciable political questions* – holding that to address climate change is the responsibility of the executive and legislative branches, not the judicial branch. To grant the plaintiffs' requested relief would, the court found, require it to balance health and environmental concerns with economic concerns to set and implement emissions caps, which involved a range of policy issues unresolved by the political branches.[38] As will become clear, dismissal on 'political question' grounds is a typical response of USA courts to climate tort lawfare.

---

37   MG Faure & A Nolkaemper *Analyses of Issues to Be Addressed: Climate Change Litigation Cases* (2007) Preface.

38   M Haritz *An Inconvenient Deliberation: The Precautionary Principle's Contribution to the Uncertainties Surrounding Climate Change Liability* (2011).

The *Kivalina v Exxon et al* case (CV 08-1138 ND Cal [2008]) concerned climate related displacement. In 2003, the USA Government Accountability Office reported that most of Alaska's native villages were affected by flooding and erosion, some of them facing imminent threats 'due in part to rising temperatures that cause protective shore ice to form later in the year, leaving the villages vulnerable to storms.'[39] Similar effects were affecting indigenous communities throughout the Arctic rendering them at risk of displacement.[40] A 2005 report by the US Army Corps of Engineers stated that the environmental situation in the Alaska Native village of Kivalina was dire and that the entire town needed immediate relocation.[41]

In this context, the Kivalina villagers, in 2008 collectively sued major USA oil companies and utilities. The narrative forcefully presented by the plaintiffs was that the oil corporations' GHG emissions contributed substantially to global warming and thus were responsible for the destruction of Kivalina by climate change related flooding. They pursued a claim of public nuisance, seeking a monetary compensation of 400 million US$, the estimated cost of relocating the village, and asked the court to order the defendants to reduce GHG emissions. Additionally, the Kivalina plaintiffs presented claims of conspiracy and concert of action against ExxonMobil, AEP, BP, Chevron, ConocoPhillips, Duke, Peabody, and Southern Company for conspiring to create a false scientific debate about climate change to mislead public opinion.[42] Like the tobacco companies were accused of hiding documents proving that tobacco was harmful, the fossil fuel industries and energy companies were blamed of conspiring to deny climate change although they had evidence otherwise.[43]

In September 2009 the Court dismissed the *Kivalina* lawsuit. As in *Connecticut v American Electric Power*, the Court found that it raised a political question unsuitable for judicial resolution and was best left to the political branches to address. To establish limits to GHG emissions and decide on compensation to affected communities was considered policy matters and left entirely to the executive. The court did not make any declarations regarding damages or rights violations of the people of Kivalina, nor regarding the conspiracy claims. The plaintiffs filed an appeal, but in 2012 appeals judges decided not to reinstate the case.

Comer v Murphy Oil (585 F 3d 855 5th Cir [2009]) was brought by landowners in Mississippi, claiming that oil and coal companies produced GHG emissions contributing to global warming, which, in turn, caused a rise in sea levels, adding to Hurricane Katrina's ferocity. The plaintiffs alleged nuisance, negligence, trespass, unjust enrichment, civil conspiracy,

39   GAO (USA) *Report of the Government Accountability Office Gao-04-142* (2003). Cited in C Shearer *Kivalina: A Climate Change Story* (2011).
40   Shearer ibid.
41   R Frank 'Kivalina and the Courts: Justice for America's First Climate Refugees?' (2011) <http://legalplanet.wordpress.com/2011/11/28/kivalina-and-the-courts-justice-for-americas-first-climate-refugees/>.
42   Ibid.
43   W Norris 'Supreme Court Deals Setback for Environmental Activists and Refugees' (2011) <http://www.praer.org/cgi-bin/mtype/mt-search.cgi?IncludeBlogs=1&tag=Maldives>.

and fraudulent misrepresentation. This case was also dismissed on political question grounds.

Beyond the USA, affected people from Nigeria – Africa's top energy producer – have filed suits against Shell in Nigerian, US American and Dutch courts. In *Gbemre v Shell Nigeria et al* (AHRLR 151, NgHC [2005]) an affected community sued Shell Nigeria for their gas flaring practice in the Niger Delta. Plaintiffs argued that gas flaring generates GHGs and subjects local communities to constant heat, light, noise, and air pollution. Over the last 40 years, pollution from flares has destroyed crops and exposed Delta residents to an increased risk of premature death, respiratory illnesses, asthma, and cancer.[44] Nigeria has been the world's biggest gas flarer, and the practice is understood to have contributed more GHGs than all other sources in sub-Saharan Africa combined. The wasteful practice also costs Nigeria about 2.5 billion US$ annually, while about 66 per cent of its population lives on less than 1 US$ a day.[45] The case thus combines (global) climate change arguments with other unwelcome (local and direct) effects of the fossil fuel industry.[46]

In November 2005 the Federal High Court of Nigeria ruled that gas flaring violates citizens' constitutional rights to life and dignity, and ordered Shell to stop the practice.[47] However, climate change arguments were not considered to sustain the decision. To date (2013), there has been no implementation of the ruling with regard to stopping the flaring practice by Shell, although given amendments made to the 1979 Associated Gas Re-Injection Act by the Associated Gas Re-injection Act Regulations of 2005, a definite date is 'soon to be set by the National Council on Niger Delta to stop gas flaring in the region' according to the Minister of Niger Delta Affairs.[48]

In *Akpan v Royal Dutch Shell/Shell Nigeria* (C/09/337050/HAZA 09-1580), a different case launched in 2008 in the Netherlands (domicile of Shell's headquarters), Nigerian farmers claimed reparations for lost income from contaminated land and waterways in the Niger. The case was linked to spills

---

44   Environmental Law Alliance Worldwide (ELAW) 'Case against Shell: Court Orders Nigerian Gas Flaring to Stop' (2012) <http://www.elaw.org/node/932>.

45   Ibid.

46   Other cases against Shell had been brought to court by Nigerians before, amongst them *Kiobel v Royal Dutch Petroleum Co* (Docket No. 10-1491, decided April 17, 2013) and *Wiwa v Royal Dutch Petroleum Co* (226 F 3d 88 2d Cir [2000]). These are lawsuits brought by the family of an Ogoni activist who was member of the Movement for the Survival of the Ogoni People (MOSOP) against Royal Dutch Shell, its subsidiary Shell Nigeria and the subsidiary's CEO, under the United States Alien Tort Statute, the Torture Victim Protection Act of 1992 and Racketeer Influenced and Corrupt Organizations Act (RICO). They are charged with complicity in human rights abuses against the Ogoni people in the Niger Delta, including summary execution, crimes against humanity, torture, inhumane treatment, arbitrary arrest, wrongful death, and assault and battery. In early June 2009, the parties announced that they had agreed to a settlement. It provides compensation for the plaintiffs and covers a portion of their legal costs. The settlement also establishes The Kiisi Trust, intended to benefit the Ogoni people.

47   For detailed case documentation see Climate Justice: enforcing climate change law 'Court Orders Nigerian Gas Flaring to Stop (14 November 2005)' <http://www.climatelaw.org/cases/country/nigeria/gasflares/2005Nov14/>.

48   E Charles 'Nigeria: Gas Flaring Will End Soon – Niger Delta Minister Vows' *All Africa* (28 April 2013). For contextual information see 'Climate change: Enforcing climate change law "Shell Fails to Obey Court Order to Stop Nigeria Flaring, Again"' (2007) <http://www.climatelaw.org/cases/country/nigeria/media/2007May2/>.

in four areas of the Niger Delta – Goi, Ogoniland, Oruma and Ikot Ada Udo—. The farmers had alleged that oil spills had poisoned their fish ponds and farmland with leaking pipelines. In January 2013 the district court in The Hague ruled Shell Nigeria, a wholly-owned subsidiary, must compensate one farmer, and dismissed four other claims filed against the Dutch parent company. The court found that the spills were not the result of a lack of security or maintenance but due to sabotage. Pursuant to Nigerian law a parent company in principle is not obliged to prevent its subsidiaries from harming third parties abroad. However, in one case, it found Shell Nigeria culpable of neglecting its duty of care and ruled that 'Shell could and should have prevented this sabotage in an easy way'.[49]

(i)   Obstacles in tort-based litigation

From a theoretical perspective a challenge for tort-based climate litigation is to *demonstrate injury in fact and a linear chain of causality*. Nuisance or damages cases point at wrongful actions that cause harm or injury to persons or their property. Damages can be caused intentionally or due to negligence. Plaintiffs of nuisance cases need to prove a) that they have suffered 'injury in fact'; b) that their injuries have been caused by the defendant, and c) that the injuries can be redressed by a court decision. Given the global nature of GHG emissions and the billions of contributors to climate change, it is not an easy case to make.[50] In climate change, there is no clear chain of causation from a particular defendant's actions to the plaintiffs' injury and a plaintiff could sue any emitter of their choosing. The lack of linear causation has been seen to count against a legal solution:

> Unlike traditional pollution cases, where discrete lines of causation can be drawn from individual polluters to their individual victims, climate change results only from the non-linear, collective impact of millions of fungible, climatically indistinguishable, and geographically dispersed emitters. [...] [C]limate change is a systematic phenomenon that is intractable to anything but a systematic political solution, one that the adversarial and insulated model of nuisance litigation is incapable of providing.[51]

Civil law based climate cases also face problems in establishing a *violation of the duty of care*.[52] Given that the extraction and use of fossil fuels has been understood as necessary and even desired in the dominant development discourses, and that many states lack regulations regarding acceptable GHG

---

49   BBC News 'Shell Nigeria Case: Court Acquits Firm on Most Charges' (2013) <http://www.bbc. co.uk/news/world-africa-21258653>.

50   Frank (note 41 above).

51   LH Tribe, JD Branson & TL Duncan *Too Hot for Courts to Handle: Fuel Temperatures, Global Warming, and the Political Question Doctrine* (2010) 15.

52   In legal literature, violation of the duty of care is generally defined as an omission to do something which a reasonable person guided upon considerations that normally regulate human affairs would do, or doing something which a reasonable person would not do. In short, it is the term used to designate a failure to exercise due care, resulting in injury to another, and for which an action for damages may be brought. Three elements have to be clear to establish a case of negligence: (1) Duty of care owed by the defendant to the plaintiff, (2) Breach of that duty, (3) Damages resulting as a breach of that duty. WW Buckland 'Breach of Duty of Care in the Tort of Negligence' (1935) 51 *L Q Rev* 637. See further D Hunter & J Salzmann 'Negligence in the Air: The Duty of Care in Climate Change Litigation' (2007) 155 *U Pa L Rev* 1741.

emissions, courts cannot easily assert that emissions by a specific company are excessive and in violation of the duty of care, hence constituting a 'negligent or unlawful act'. A comprehensive scientific understanding of the adverse effects of fossil fuels-dependent industry in the global climate (and on environmental, socio-economic and cultural circumstances) is relatively new. On-going debates on emissions standards and regulations may transform the way in which the duty of care and its violation are understood in this context.

These challenges notwithstanding, the discussion above shows that, particularly in the USA, *the main obstacle to civil suits in the context of climate change has been the political question doctrine.*[53] In climate change cases, judges have found that by deciding on GHG cuts they would be formulating emissions policy, which would mean engaging in political judgments reserved for congress and the executive. District courts would have to fix and impose future emission standards upon defendants and all other emitters, and judges would be providing 'government by injunction' which would not be a product of the consent of the governed. In this sense, the needed regulations to mitigate climate change are understood to be political rather that justiciable.

So, given the lack of court victories, why do lawyers continue to pursue these cases? Some lawyers think that with continuing pressure, climate tort lawfare will at some point present real opportunities for successful litigation.[54] But there are also potentially indirect gains. The threat of liability from on-going court cases, could lead industry and others to promote a liability regime under the UNFCCC that would clarify the rules of liability and essentially limit private sector obligations – as has been done previously with environmental damage from nuclear facilities and oil spills.[55] And although court actions fail, litigation is also used for oblique proposes, such as raising public awareness or mobilize public opinion, keeping the issue on the political agenda or building pressure for legislative and policy action, creating influence to supplement other strategies and pressing the opposition to settle.[56] Dismissal of cases on political question grounds in the USA has put the public eye on the political

---

53 Constitutional Law distinguishes between on the one hand, 'legislative' and 'executive' questions, which require pluralistic processes of legislation and treaty making and are reserved for the political branches, and 'cases' and 'controversies' reserved for judicial resolution. For various accounts on the political question doctrine, see for instance E Chemerinsky 'Who Should Be the Authoritative Interpreter of the Constitution? Why There Should Not Be a Political Question Doctrine' in N Mourtada-Sabbah & BE Cain (eds) *The Political Question Doctrine and the Supreme Court of the United States* (2007) 181-98; A Thorpe 'Tort-Based Climate Change Litigation and the Political Question Doctrine' (2008) 24 *Journal of Land Use* 79; Tribe et al ibid.

54 See for instance D Grossman 'Warming up to a Not-So-Radical Idea: Tort-Based Climate Change Litigation' (2003) 28 *Colum J Envtl L* 9. See also Hunter (note 1 above).

55 See for instance Brussels Convention Relating to Civil Liability in the Field of Maritime Carriage of Nuclear Material (1971); Convention on Civil Liability for Oil Pollution Damage Resulting from Exploration for and Exploitation of Seabed Mineral Resources, November 1977 (1977); Int'l Convention on Civil Liability for Oil Pollution (1969); Paris Convention on Third Party Liability in the Field of Nuclear Energy (1960); Protocol of 1992 to the International Convention on the Establishment of an International Fund for Compensation for Oil Pollution (1971); Vienna Convention on Civil Liability for Nuclear Damage (1963). Cited in Hunter (note 1 above).

56 Lin (note 24 above).

branches for solutions.[57] But there are also court cases directly addressing the political branches.

(b) Climate regulation lawfare: unblocking governance through administrative law claims

Individuals, NGO's and subnational government institutions have used *judicial review actions*[58] among other to challenge States' failure to curb GHG emissions and conduct environmental impact assessments on projects they finance or permit.

In *Massachusetts v Environmental Protection Agency (EPA)* (549 US 497 [2007]), twelve USA States and several cities brought a lawsuit against the EPA to make it to regulate CO_2 from vehicles and other GHGs as air pollutants. In 2003 EPA stated that the Clean Air Act did not grant the agency authority to regulate CO_2 emissions from vehicles and other GHGs for climate change purposes – and even if it had been granted such authority, EPA would decline to set GHG emissions standards for new vehicles. In this case, the USA Supreme Court made its first pronouncement on global climate change, and decided 5-4 in favour of the plaintiffs, accepting provided evidence on the fact that global warming exists and that emissions from vehicles can have impacts on citizen's health and the Massachusetts coastline. Hence the Court ordered EPA to include GHGs as air pollutants and regulate emission standards for motor vehicles. Later the EPA stated that six greenhouse gases including CO_2 endangered public health and were likely responsible for the global warming experienced over the past half century. This case – as the first USA Supreme Court acknowledgement of the link between fossil-fuels usage and environmental harm, of the pressing need for regulation of GHG emissions, and of the role of courts in unblocking regulatory processes – has had immense importance for USA environmental advocacy groups and for transnational litigation.

But *Massachusetts v EPA* was not the first case with this type of claims. In the USA, the Minnesota Court of Appeals in *The Matter of Quantification of Environmental Costs* (578 NW 2d 794 Minn App [1998]) helped establish global warming as a harm that could be redressed by state-level legislative and administrative action. The regulation and following litigation fostered political voice, increased pressure for reform, and encouraged utilities to voluntarily cut pollution through the warning of future adverse regulatory decisions.[59] A New Zealand case, *Environmental Defense Society v Auckland Regional Council and Contact Energy Limited* (NZRMA 492 [2002]) was one

---

57 Hunter (note 1 above).
58 Judicial review of administrative decisions is a legal action intended to ensure that powers are exercised for the purpose they are conferred for, and in ways in which they are intended to be exercised. JJ Spigelman 'The Integrity Branch of Government' (2004) 78 *Australian Law Journal* 724, 730. Cited in BJ Preston 'The Role of Courts in Relation to Adaptation to Climate Change' (2008) *Adapting to climate change: Law and policy conference.*
59 S Stern 'State Action as Political Voice in Climate Change Policy: A Case of the Minnesota Environmental Cost Valuation Regulation' in Burns & Osofsky (note 34 above). See further for a comprehensive account on this case.

of the first cases of judicial acknowledgement of climate change science and human influence.[60]

In *Natural Resources Defense Council et al v Abraham et al* (355 F 3d 179, 2d Cir [2004]), various NGOs and subnational USA States sued the Department of Energy (DOE) for weakening energy efficiency standards. The Energy Policy and Conservation Act required the DOE to set energy efficiency standards for appliances at the maximum level that is technologically and economically feasible. In early 2001, the new Bush government sought to weaken DOE standards for air conditioners. The lawsuit was based on the global warming impact of the increased emissions. The ruling by the USA Court of Appeals for the Second Circuit held that the weakening of the standards violated the Energy Policy and Conservation Act, consequently the court reinstated the stronger standards. The decision was the result of a lawsuit filed by 10 states, the National Resources Defence Council, the Consumer Federation of America, other consumer groups, and state utility regulators.[61]

The European Court of Justice handed down a ruling in December 2011 for the case *Air Transport Association of America et al v Secretary of State for Energy and Climate Change* (Case C-366/10), relating to the European Union Emissions Trading Scheme (EU-ETS). A lawsuit by USA air carriers that challenged the EU Directive 2008/101 for including all non-EU carriers into the EU-ETS. The Court ruled that airlines based outside the European Union must abide by contentious EU legislation requiring them to pay for their carbon pollution.[62]

The various cases regarding the EU-ETS 'exemplify the courts' involvement in climate change governance by ensuring that the objectives of the regulatory response are met'.[63] The stated objectives of the EU-ETS are the reduction of GHG gases, for the EU to be able to comply with their Kyoto obligations. However, in the actual cases[64] emphasis seems to be placed mostly on economic considerations rather than on the environmental objectives of the EU-ETS.[65] The narratives playing out in EU-ETS court cases focus on the formal breach of contractual commitments (Kyoto) and the maintenance of economically efficient conditions of the carbon trade scheme, the preservation of the current European development model and employment, and the defence of the integrity and conditions of competition of internal markets, all in the

---

60   Greenpeace 'Greenpeace Briefeing: History of Climate Change Litigation' (2007) <http://www.greenpeace.org/new-zealand/PageFiles/113249/history-climate-change-litigation.pdf>.

61   For an overview of subsequent DOE rulings on energy eficiency standards after this and other court cases, see Department of Energy, Docket Number EERE-2011-BT-STD-0011 <http://www1.eere.energy.gov/buildings/appliance_standards/residential/pdfs/cacfurn_dfr_final-version.pdf>. For background information on a similar case in California see Earth Justice 'California Ag, Environmental Groups in Court over Weak Energy Efficiency Standards' (2009) <http://goo.gl/smbxw>.

62   J Chaffin & A Parker 'Foreign Carriers Must Pay EU Carbon Fees' *The Financial Times* (22 December 2011).

63   Lin (note 24 above) 49.

64   See for instance *Germany v Commission* (Case T374/04) and *Poland v Commission* (Case T-183/07). Cited in Lin ibid 49.

65   Lin ibid.

context of complying with emissions' reduction as contractual obligation of the EU.[66]

A range of court cases have demanded climate responsibility in industrial investments, focusing on the damage to humans, including future generations, and demanding environmental impact assessment specific to climate change. Australian courts in particular have frequently been engaged in such cases. Australia has large resource of coal,[67] a consolidated energy industry, already evident coastal erosion problems, and a number of lawsuits have challenged lack of regulation of environmental impact assessment of new coalmines and coal-fired power plants. An important case among these is *Gray v Minister for Planning* (NSWLEC 720 [2006]), which challenged an open cut coalmine permit for lack of environmental impact assessment specific to climate change. The Land and Environment Court of New South Wales held that GHG emissions resulting from burning of coal extracted from the projected mine should be considered in the environmental assessment. The judgment was significant in stating that consideration of the principles of environmentally sustainable development and particularly in the principle of *intergenerational equity* and the *precautionary principle*, meaning that emissions should have been included in the environmental assessment.[68] Subsequent to this ruling, the New South Wales government introduced a policy in 2007, to guarantee that indirect emissions from new extractive industries were considered in the decision-making of licensing and permitting process.[69]

Several cases in Europe and the USA have argued for climate change concerns to be taken into account when public money is invested abroad. In *Germanwatch and BUND v Ministry of Economics and Labour* (VG 10 A 215.04 [2006]), German NGOs sued the Export Credit Support Agency for not disclosing information on the GHG emissions of the projects they support. Allegations were that the export credit agency was secretly funding fossil fuel projects abroad, which were relevant to the Kyoto Protocol. The German Administrative Court ruled that 'the primary purpose of an export credit is to support the German economy and not environmental protection' however, the court supported the plaintiffs' argument that financing overseas projects contributing to climate change was contrary to obligations under the Kyoto protocol. It ruled that when granting export credit for energy production projects, this would constitute a measure and activity that would likely affect the environment, and information in this regard could not be denied.[70]

Along a similar line, *Friends of the Earth et al v Watson* (2005 WL 2035596, 35 ENVTL L REP 20, 179) was filed in the USA by NGOs and

---

66   Ibid.
67   Faure & Nolkaemper (note 37 above).
68   BJ Preston 'The Influence of Climate Change Litigation on Governments and the Private Sector' (2011) 2 *Climate Law* 485. For a comprehensive account of Climate change litigation in Australia, see ibid; Preston (note 58 above).
69   Preston 'The Influence' ibid. See further for an account on environmental impact assessment cases in Australia.
70   See *Bund für Umwelt und Naturschutz Deutschland EV & Germanwatch EV v Federal Republic of Germany represented by the Minister of Economy and Labour* (BMWA), Administrative Court Berlin. BUND is the German section of NGO Friends of the Earth.

four cities against Export-Import Bank and the Overseas Private Investment Corporation under the National Environmental Policy Act (NEPA). They contested the provision of billions of dollars for the finance and insurance of oil fields, pipelines and coal-fired plants in developing countries over the previous 10 years, without assessing impacts on the environment including global warming. In 2009 the parties reached a settlement favourable to the plaintiffs.[71]

Before the UK Administrative Court - High Courts of Justice, plaintiffs in *World Development Movement et al v UK Treasury* (EWHC 3020 (Admin) [2009]) aimed to provide judges with evidence of the harmful investments being made by the Royal Bank of Scotland (RBS) significantly increasing the level of CO_2 emissions and resulting in the abuse of human rights. The High Court denied the request for permission to hold a judicial review over the Treasury's actions.[72]

Climate responsible investments have also been brought up in quasi-judicial bodies referring to corporations' violations of the guidelines of the Organization for Economic Co-operation and Development (OECD). Cases such as *Germanwatch v Volkswagen* (2007) and *Norwegian Climate Network et al v Statoil* (2011) alleged that corporations breached OECD Guidelines by irresponsibly investing in fossil fuel related economic activity contributing to climate change. Germanwatch argued that Volkswagen had a duty to adopt business strategies avoiding climate change and to implement climate protection goals.[73] The Norwegian Climate Network and Concerned Scientists Norway alleged that Statoil ASA had breached the OECD Guidelines by investing in the oil sands of Alberta, Canada, and thereby contributing to Canada's violation of international Kyoto obligations to reduce GHG emissions in the period 2008-2012. Requesters argued for leaving oil sands unexploited to stabilize global GHG emissions.[74] Both cases were rejected as being beyond the scope of the OECD Guidelines.

Administrative law claims have also targeted urban planning for failing to plan for CO_2 emissions management. In *California and San Bernardino County* (Superior Court Case No. CIVSS 700329 USA [2007]) litigants demonstrated

---

71   The Bank can continue funding fossil fuel projects, but must disclose the estimated CO_2 emissions from potential transactions for such projects; make public its decisions as to whether the NEPA applies to specific transactions involving fossil fuel projects and provide opportunity for comment; develop and implement a carbon policy in cooperation with plaintiff representatives that will include a renewable energy loan facility; and promote consideration of climate change issues among export credit agencies. See S Ilias *Export-Import Bank: Background and Legislative Issues* (2011).

72   For the judgment in this case, see <http://peopleandplanet.org/dl/ddd/justicesalesruling. pdf>. For an account from the plaintiffs perspective, see for instance Platform London 'The Treasury, UKFI and RBS – the Government's Biggest Climate Change Failure.' (2010) <http:// platformlondon.org/p-article/treasury-ukfi-rbs-climate-change-failure/>. For a summary of allegations by all the parties see High Courts of Justice 'First Witness Statement, Co/5323/2009' (2009) <http://peopleandplanet.org/dl/ddd/legalchallenge/witnessstatement.pdf>.

73   OECD Watch 'Germanwatch v Volkswagen' <http://oecdwatch.org/cases/Case_119>.

74   OECD Watch 'Norwegian Climate Network et al v Statoil' <http://oecdwatch.org/cases/ Case_248>.

how 'urban growth plans significantly impact emissions trajectories.'[75] Part of the legal argument is that even though individual's choices have a minor impact on total worldwide emissions, personal choices matter as they add up.[76] Individuals can decide on their own carbon footprint by deciding on types of goods that they prefer to consume or choosing amongst various transport options. However, individual decisions are influenced by socio-cultural and multiscalar legal contexts.[77] Studies suggest urban planning practices such as zoning, the use of large individual lots, separation between residential and commercial uses, and limited public transportation influence vehicle usage and kilometres travelled and the overall emissions from a locality. This case ended in a settlement by the involved parties.[78]

A growing area of climate-related litigation concerns harms arising from climate mitigation efforts, and, particularly in Australia, from windfarm developments. A typical such case is *Taralga Landscape Guardians Inc v Minister for Planning* (NSWLEC 59 [2007]). Cases are brought by local community members concerned with the amenity, landscape, and potential health effects of wind-farm expansions. Decisions have been mixed and reflect diverging opinions on wind-farming as climate mitigation. Current precedent is that while the 'broader public good' associated with renewable energy development and climate change mitigation should be given due consideration, it will not always outweigh the amenity impacts of wind turbines on surrounding properties.[79]

Legal action has also been used to challenge government failure to adapt to climate change, and to take steps to avoid major and preventable social harm. In Argentina, after the 2003 Santa Fe floods, which caused major damage and loss of lives, citizens successfully used Article 6 of the UNFCCC and the *'Acción Informativa'* mechanism to expose failure to adapt to climate change. According to Article 6 of the UNFCCC, in carrying out their commitments State Parties shall promote and facilitate public access to information on climate change and its effects. The legal action revealed that while the authorities had made plans for infrastructure changes needed to protect people, these had not been acted upon.[80]

As this discussion shows, administrative law based climate lawfare is wide-ranging, from compelling government agencies to regulate GHGs, and assess climate impacts at the project level, to demands that climate change

75   Osofsky (note 26 above) 593 Fn 24.
76   Ibid.
77   Ibid.
78   See 'Confidential Settlement Agreement *People v County of San Bernardino*' (2007) <http://ag.ca.gov/cms_pdfs/press/2007-08-21_San_Bernardino_settlement_agreement.pdf>. Cited in Osofsky (note 26 above) 593 Fn 24.
79   University of Melbourne 'Climate Change Law, Cases Challenging Windfarms' (2011) <http://blogs.unimelb.edu.au/peel_climatechange/2011/09/01/cases-challenging-windfarms/>.
80   The case was concluded August 7, 2006 (Administrative Record file number 1200253510000651041, National Secretariat of Environment and Sustainable Development, Argentina. English translation on file with the authors). See also Greenpeace Brifing (Climate, New Zealand, June 2007) 'History of climate change litigation' <http://www.greenpeace.org/new-zealand/PageFiles/113249/history-climate-change-litigation.pdf>.

considerations be incorporated into public financing and urban planning decisions. Results are mixed, but these collective and forward-looking measures avoid some of the problems hampering tort-based claims, and have generally been more successful in court. They have established important principles such as the recognition of anthropogenic climate change; the precautionary principle and obligations towards future generations.

While USA courts in particular, have carefully avoided making any declarations on possible links between climate change and the fossil fuel industry in civil law cases, judgments in administrative law cases have frequently made use of scientific arguments *on anthropogenic global warming*, and elaborated on the *precautionary principle* as means to enforce action from the political branches on regulation matters. This is of great political significance. When courts base their decisions in scientific arguments on human induced climate change, the general public's attention shifts from whether climate change is occurring to what the appropriate responses should be.[81]

When the USA Supreme Court, in *Massachusetts v EPA*, ruled that global warming existed, was human made, and caused loss of shoreline, EPA had not contested the links between GHG emissions and climate change. Its decision not to regulate GHG emissions from new motor vehicles was based on their alleged *insignificant* contribution to air pollution and climate change. In an important piece of jurisprudence on climate litigation, the USA Supreme Court decided that small, incremental steps can be attacked in a federal judicial forum, because public agencies – like legislatures – do not generally resolve massive problems in one fell regulatory swoop.[82] The Court said that reducing domestic automobile emissions (a major contributor to GHG concentrations) was 'hardly a tentative step.'[83]

Judicial review actions addressing (lack of) climate change regulations, presents climate change fundamentally as a governance challenge and litigation as a medium in which to 'debate the appropriateness and necessity of regulatory entities at different scales taking particular steps to address global climate change.'[84] In sum, adjudication is here interpreted as 'a form of climate change governance.'[85]

The precautionary principle in the context of environmental protection relates to the management of scientific risk. It is a fundamental component of the concept of ecologically sustainable development and is defined in Principle 15 of the Rio Declaration (1992), 'Where there are threats of serious or irreversible environmental damage, lack of full scientific certainty should

---

81   Hunter (note 1 above).
82   Preston 'The Influence' (note 68 above).
83   *Massachusetts v EPA* 549 US 497, 127 S Ct 1438, 167 L Ed 2d 248 (2007) 21-22. Cited by Preston ibid 4.
84   D Markell & JB Ruhl 'An Empirical Assessment of Climate Change in the Courts: A New Jurisprudence or Business as Usual?' (2012) 64 *Fla L Rev* 16. In the same sense see also Osofsky (note 26 above).
85   Lin (note 24 above) 35.

not be used as a reason for postponing measures to prevent environmental degradation.'[86]

As noted above, various Australian cases have elaborated on the precautionary principle, especially in relation to the licensing of new coalmines and urban coastal projects.[87] By using existing regulations to require decision-makers to consider future climate risks in planning decisions, some cases have led to the revision or formulation of public policies on mining and coastal management. Less successful cases have nonetheless underlined areas in need of law reform. Recent high profile cases targeting major sources of GHG emissions in Australia have raised innovative arguments based on common law public nuisance grounds and the public trust doctrine.[88] Further jurisprudential development of the precautionary principle in the frame of climate change could lead to advances in executive branches' reduction of emissions, more careful practices in licensing of fossil fuel related projects, or work in favour of financing of alternative energy research and projects.

(c)   Transboundary climate lawfare: seeking climate justice thought public international law

Given that climate change has a truly global reach surpassing the capability of any country, individually considered, to fully address it,[89] and current global regulatory efforts have proven impotent, activists have turned to international law for remedies. There is no designated international court or treaty body for environmental law, but complaints have been placed before different international quasi-judicial bodies, drawing on various branches of law, in particular international human rights law. The cases concern the adverse effects of climate change on indigenous peoples, and on communities living in non-industrialized countries with little access to adaptation assistance and therefore most exposed to climate harms. There are also successful claims regarding special protection of natural sites considered part of the world's heritage by the United Nations Educational, Scientific and Cultural Organization (UNESCO).[90]

The two cases discussed below have drawn particular attention:

In 2005, an NGO consisting of the Inuit people of Alaska, Canada, Greenland (Denmark) and Russia – the Inuit Circumpolar Conference – filed a petition to the Inter-American Commission on Human Rights, alleging that the USA was violating the human rights of Arctic people by refusing to limit GHG emissions. The petition was supported by scientific studies finding that global warming affects weather patterns, thins the ice and hence impedes the

---

86   D Cole *The Precautionary Principle – Its Origins and Role in Environmental Law* (2005) Unpublished paper <http://www.laca.org.au/images/stories/david_cole_on__precautionary_ principle_EDO.pdf>.

87   See for instance K Mcgree 'The Legal Obligation to Consider Greenhouse Gas Emissions in an Application of the Precautionary Principle' (2009) 2 *Queensland Law Student Review* 45.

88   Preston 'The Influence' (note 68 above).

89   Gerrard (note 36 above).

90   See UNESCO 'Case Studies on Climate Change and World Heritage' (2007) <http://whc.unesco. org/en/news/545/>.

Intuits' ability to live off their polar environment.[91] It was based on the claim that the USA was one of the world's largest producers of carbon emissions per capita.[92] Petitioners alleged violations including of ESC rights 'to enjoy the benefits of culture', 'to use and enjoy lands traditionally occupied', 'to use and enjoy personal property', 'to the preservation of health', 'to life, physical integrity and security', 'to their own means of subsistence', and 'to residence, movement and inviolability of the home'.[93]

The narrative of the *Inuit Circumpolar Conference* case tells a particular story of climate change centring on the argument that by not reducing GHG emissions when their adverse effects in the artic are increasingly known and scientifically understood, highly emitting countries are putting at risk and violating the ESC rights of Inuit peoples. The petition asked the Inter-American Commission to make recommendations to the USA to reduce its emissions and create plans to protect Inuit culture and resources through adaptation assistance.[94] The case was rejected one year later by the Commission arguing that it lacked evidence of concrete harms and failed to establish how the alleged facts characterized a violation of rights.[95]

A similar narrative was presented by the Maldives, one of the small island States of the Indian Ocean, which in 2008 submitted a petition to the Office of the High Commissioner for Human Rights (OHCHR), under Human Rights Council Resolution 7/23 'Human rights and climate change.'[96] The Maldives stressed 'its nation's reliance on the physical integrity of its island home', which is threatened by rising sea levels allegedly connected to climate change, and which is a result of 'social processes beyond its sovereign control.' The Maldives used the logic of 'territorial sovereignty' to claim extra-territorial obligations from the international community.[97] Like *Kivalina v Exxon*, the Maldives case adds to the claims of groups likely to become 'environmental refugees' as some are framing it.[98] Part of their argument is that many people are currently being displaced by climate change 'but a person whose island is swallowed up by rising sea level has

91 University of Melbourne 'Climate Change Law, Australian and Overseas Developments: Inuit Petition to the Inter-American Commission on Human Rights' (2010) <http://blogs.unimelb.edu.au/peel_climatechange/2010/05/24/inuit-petition-to-the-inter-american-commission-on-human-rights/>.

92 Frank (note 41 above).

93 See Inuit Circumpolar Conference *Petition to the Inter-American Commission on Human Rights Seeking Relief from Violations Resulting from Global Warming Caused by Acts and Omissions of the United States* (2005) <http://www.inuitcircumpolar.com/files/uploads/icc-files/FINALPetitionICC.pdf>.

94 Shearer (note 39 above).

95 University of Melbourne (note 91 above).

96 Maldives *Submission of the Maldives to the OHCHR under Human Rights Council Res. 7/23* (2008). See Resolution 7/23: Human Rights and Climate Change (2008). Assessment at national level of the impact of climate change (experienced or anticipated) on human lives and on population most affected and vulnerable.

97 See Inuit Circumpolar Conference *Petition to the Inter-American Commission* (note 9 above).

98 See for instance DC Bates 'Environmental Refugees? Classifying Human Migrations Caused by Environmental Change' (May 2002) 23 *Population and Environment* 465; S Castles 'Environmental Change and Forced Migration: Making Sense of the Debate' (2002) *New Issues In Refugee Research, Working Paper No. 70.*

no rights under Refugee law, in fact, they are considered 'migrants' meaning they are 'voluntarily' leaving their country.'[99]

The underlying narrative of the small island states affected by rising sea levels and whose inhabitants may need relocation is that industrialized nations are causing transnational climate-related problems; therefore they should pay the costs and lead the way to reform.[100] The UN Development Program estimated that, by 2015, industrialized nations must provide around 86 billion US$ per year to assist people most vulnerable to ruinous floods, droughts and other events that scientists think will accompany global warming.[101]

So far Public International Law petitions have not resulted in judicial orders or recommendations to States about changing/diversifying their energy sources or improving energy efficiency, neither have they resulted in any declarations on human rights violations or protection for communities claiming special vulnerability for climate change. Hence, in terms of direct effects not much has been achieved.

On the other hand, these and other international cases have clearly demonstrated –and drawn wide attention to– the urgent problems of trans-boundary equity that climate change raises, with those who have contributed the least to GHGs accumulation being the ones having more trouble financing adaptation plans. They also bring out complexities arising from different countries and population group having diverging adaptation priorities.[102] Hence, despite the limited legal successes, this is an area of red-green lawfare that advocacy groups in the global south are likely to use more actively in the coming years.

III   EMERGING LITIGATION STRATEGIES

Litigation is always based on the alleged breach of an obligation, engaging the responsibility of a wrongdoing party.[103] Climate litigation conveys narratives in terms of who is causing a problem, who is suffering the consequences and therefore should be compensated, and who should take which action to implement climate change mitigation and adaptation measures. We have now seen how, since the early 2000's litigants have pursued different strategies, mostly within administrative law, civil law, and international human rights law. Litigation strategies are constantly changing, but based on the current state of legal affairs there are potentially promising avenues for litigation

---

99 W Norris 'Supreme Court Deals Setback for Environmental Activists and Refugees' (2011) <http://www.praer.org/cgi-bin/mtype/mt-search.cgi?IncludeBlogs=1&tag=Maldives>. According to the definition by UN High Commissioner on Refugees, (UNHCR) a Refugee is someone either inside or outside their national borders fleeing persecution due to their affiliation with a social group – for instance ethic, religious, political etc. Some contest the notion of climate refugees arguing that is leads to decreasing attention in the problem of political persecution.

100 Ibid.

101 UNDP *Human Development Report: 'Fighting Climate Change, Human Solidarity in a Divided World'* (2007-2008).

102 Lin (note 24 above).

103 MG Faure & A Nollkaemper 'International Liability as an Instrument to Prevent and Compensate for Climate Change' (2007) 26 *Stan Envtl L J* 123.

that include some specific public law strategies aiming at the improvement of climate mitigation and adaptation policies.

One emerging strategy further explores the concept of **intergenerational responsibility**, positively accepted by the Supreme Court of Philippines in *Oposa v Factoran* (224 SCRA 792 [1993]). In this case the petitioners, all minors, asked the Department of Environment and Natural Resources to cancel all timber licenses due to deforestation. The Court accepted the children's standing to ask for the protection of forests for future generations. However, no implementation has followed the celebrated court decision.[104]

Similar climate change cases in the name of children have been filed against the USA government under *the public trust doctrine*, alleging breach of the government's duty to regulate GHGs.[105] This effort is known as *atmospheric trust litigation*. The cases ask for the protection of the rights of future generations and recall the necessity to develop innovative legal responses to address the carbon loading of the atmosphere. The plaintiffs have sought declaratory and injunctive (preventive) relief requiring governments to reduce CO_2 emissions by at least six per cent per year starting in 2013. So far, the USA judiciary has dismissed the cases as a matter of jurisdiction, holding that public trust cases are not federal (constitutional) but a creation of state law. Judges have also argued that even in the case that the protection of the atmosphere were to be consider a constitutional matter, it would be displaced by the Clean Air Act with respect to regulation of CO_2.[106]

Australian Judge Brian J Preston, comments that various high-profile cases in the USA and Australia have discussed the extension of current common law principles to the problem of climate change. If, for instance, 'the law of public nuisance is extended to the impacts of climate change or the public trust doctrine is extended to include an atmospheric trust, courts could potentially hold corporations liable for their GHG emissions and force governments to regulate emissions rather than merely requiring decision-makers to consider the climate change impacts of individual developments (as most of the previous climate change cases have done).'[107] Even if the applicants do not succeed in court, these cases could contribute to the evolution of climate change jurisprudence, and put pressure on legislators and industry to reduce GHG emissions.[108]

Deforestation could also become central to these cases. The Philippine organization People's Network for Environment are considering going to court to claim that the government should not approve mining projects, under

---

104 Child Rights International Network 'Minors Oposa v Secretary of the Department of Environmental and Natural Resources' (2013) <http://www.crin.org/Law/instrument. asp?InstID=1260>.

105 See for instance *Alec L v Lisa Jackson et al* (Civil Action No. 11-cv-2235 (RLW); K Ellison 'An Inconvenient Lawsuit: Teenagers Take Global Warming to the Courts' (2012) <http://www. motherjones.com/environment/2012/05/alec-loorz-global-warming-lawsuit>.

106 S Balanson 'Federal District Court Dismisses Climate Change "Public Trust" Lawsuit' (2013) <http://www.lexology.com/library/detail.aspx?g=b6688961-ac09-41b2-9819-298621e68194>.

107 Preston 'The Influence' (note 68 above) 20.

108 Ibid.

the argument that mining has led to massive deforestation, destroying CO_2 sinks around the entire country.

Another emerging avenue for climate litigation is based on the **rights of nature**, which emerges from the principle that humans do not have the right to destroy the natural environment. Ecuador and Bolivia amended their constitutions after pressure from the large indigenous populations who place the environment and the earth at the centre of all life. But despite the constitutional recognition of the rights of mother earth, it remains to be seen whether and how this will translate into policy. Neither Ecuador nor Bolivia have made decisive advances in addressing the serious environmental and social problems caused by mining, and oil and gas extraction. Oil companies continue to undertake deforestation processes, amidst inertia in governmental policies –caused in part by the tensions between widely diverging narratives– (and interests) on development and conservation.

The concept of **wild law** has also gained power in recent years. Under wild law, natural ecosystems' rights surpass the interests of any one species, including humans. Numerous conferences have deliberated how to apply wild law to climate change mitigation efforts.[109]

Both conceptions of the rights of nature could give rise to new forms of climate change lawfare domestically as well as internationally.

Litigation may also emerge in relation to **transnational mitigation efforts**, such as REDD+.[110] Critics have feared the social effects of REDD+, arguing that companies with vested interests in REDD+ are corporations involved in monoculture plantations, large-scale agribusiness, industrial logging, hybrid seeds, extractive industries, and financial investment banks.[111] Civil society organizations and grassroots movements in the global south already express concern towards REDD+, citing among other, insufficient consultation processes with local communities; criteria to determine when a country is ready to implement REDD+; potential negative effects such as loss of biodiversity due to hurried agreements and lack of planning; lack of safeguards to protect Indigenous Peoples' rights to land and self-determination; lack of regional policies to stop deforestation, and dispossession of indigenous peoples, small farmers, traditional communities and forest-dependent peoples. Most REDD+ projects are still in the design and negotiation phases, but as projects materialise, local adverse effects of the mechanism could start to be debated in courtrooms, combining red and green issues.

---

109 C Mellino 'Bolivia and Ecuador Grant Equal Rights to Nature: Is "Wild Law" a Climate Solution?' (2011) <http://thinkprogress.org/climate/2011/11/21/373273/bolivia-and-ecuador-equal-rights-to-nature-wild-law-climate-solution/?mobile=nc>.

110 Redd+ is a climate change mitigation strategy aiming, through market mechanisms and financial incentives, to prevent deforestation and forest degradation in developing countries in order to reduce global emissions of GHGs.

111 See for instance C Lang '"We Reject Redd+ in All Its Versions" – Letter from Chiapas, Mexico Opposing Redd in California's Global Warming Solutions Act (Ab 32)' (2013) <http://goo.gl/d2YLG>; Global Justice Ecology Project *No Redd Papers Vol 1* (2011); J Cabello & T Gilbertson *No Redd: Una Lectura Crítica* (2010) Collection of Articles written by various organizations <http://noredd.makenoise.org/wp-content/uploads/2010/REDDreaderES.pdf>.

There is also a growing discussion in regard to the use of **criminal law** to penalize environmental harms flowing from global warming, including efforts to make "ecocide" a concept under which global warming and climate change would become not just an environmental problem but an international crime.[112] Moreover, there is a growing discussion on the possibility of states defining a maximum allowed contribution to the greenhouse effect and categorizing excessive GHG emissions and/or deforestation by corporations as a crime in their domestic legislation.[113]

As it is increasingly clear that harmful climate change cannot be avoided, increasing attention is focused on adaptation. The international normative framework for climate change puts some pressure on governments to create and implement prevention and emergency aid assistance plans. If such plans do not exist, the less privileged communities – often not covered by insurance, and with little or no access to ordinary healthcare, housing programs, and social security – have greater challenges averting and overcoming phenomena like flooding, storms, rising sea levels, and loss of land for coastal erosion.

The growing scientific input on harms from climate change can be used in combination with the UNFCCC and its Kyoto protocol; the responsibility to protect principle;[114] or in combination with legal norms on economic, social and cultural rights, to pursue **judicial enforcement of socio-economic rights** in the context of assistance and protection for climate related vulnerability.[115] With the jurisprudence emerging on socio-economic rights, particularly in the global south, this appears to be a promising path.[116]

One of the most important lessons from ESC rights litigation has been the realizations that not only are these rights justiciable, but also when carefully designed and implemented, rulings can contribute to the transformation of

---

112 See MA Gray 'The International Crime of Ecocide' (1995-1996) 26 *Cal W Int'l L J* 215; P Higgins *Eradicating Ecocide* (2010).

113 See for instance RC Kramer & RJ Michalowski 'Is Global Warming a State-Corporate Crime?' in R White (ed) *Climate Change from a Criminological Perspective* (2012) 71-88.

114 For a discussion on the use of responsibility to protect beyond the human security debate, see for instance G Evans *The Responsibility to Protect in Environmental Emergencies* (2009); LA Malone 'Green Helmets: Eco-Intervention in the Twentyfirst Century' (2009) 103 *Am Soc'y Int'l Proc* 19.

115 Some efforts in this direction are under way. For instance, Global Legal Action on Climate Change (GLACC), a Philippine civil society organization has asked the High Tribunal to compel the government to implement two existing laws that could ease flooding which is are a common problem in the Philippines and expected to worsen as climate change brings more severe storms and sea level rise. PS Romero 'Philippines: Landmark Court Challenge Could Force Action to Curb Climate-Related Flooding' (2010) <http://reliefweb.int/report/philippines/philippines-landmark-court-challenge-could-force-action-curb-climate-related>.

116 See for instance R Gargarella, P Domingo & T Roux *Courts and Social Transformation in New Democracies: An Institutional Voice for the Poor?* (2006). V Gauri & DM Brinks *Courting Social Justice: Judicial Enforcement of Social and Economic Rights in the Developing World* (2008); M Langford *Social Rights Jurisprudence: Emerging Trends in International and Comparative Law* (2008); S Liebenberg *Socio-Economic Rights: Adjudication under a Transformative Constitution* (2010); C Mbazira *Litigating Socio-Economic Rights in South Africa: A Choice between Corrective and Distributive Justice* (2009); C Rodriguez-Garavito 'Beyond the Courtroom: The Impact of Judicial Activism on Socioeconomic Rights in Latin America' (2011) 89 *Tex Law Rev* 1669; A Yamin *Los Derechos Económicos, Sociales Y Culturales En América Latina: Del Invento a La Herramienta* (2006); A Yamin & S Gloppen (eds) *Litigating Health Rights* (2011).

structural social problems.[117] A variety of court cases on socio-economic rights, including in Latin America, South Africa, India and Nepal, have addressed issues of inequality and special vulnerability of marginalised communities. Unlike USA courts, which have carefully applied the political question doctrine in ESC rights cases, courts elsewhere have developed a different jurisprudence.

When confronted with public nuisance or ESC rights violations, some global south courts have been more inclined to order the political branches to design a public policy that is respectful of and advancing those rights. Decisions have included declaration of rights, design of remedies, and monitoring mechanisms, including creation of consulting commissions to observe their implementation.[118] Since remedies to structural socio-economic problems cannot be disconnected from public policy, some courts have given orders to the political branches to create plans and negotiate remedies with civil society, industry, or petitioners' coalitions.[119] In this way courts have found a role in pro of the protection of social rights without having to encroach on the essential functions of the political branches. While judges may lack technical expertise to create or re-design complex policies, courts may –by declaring the nuisance or rights violation and calling for the political branches to act– help unblock administrative inertia and coordinate societal efforts to overcome a governance problem.[120]

Litigating climate change in much closer connection to socio-economic and cultural rights, and taking advantage of the jurisprudential developments in this field in global south countries, could lead to a better integration of red and green lawfare in climate change contexts.

The need for **transnational and multiscalar solutions** –both for climate mitigation and adaptation– is increasingly clear. Many court cases show that the division of responsibilities concerning climate change is blurred between private actors and public authorities at subnational, national and international level. Litigation contributes to the governance debate by giving weight to and managing the different interests involved, restating responsibilities and emphasizing what are the different interests to take into account when deciding upon what should be done by whom.[121] International public law cases like the *ICC Inuit petition* contribute to fostering dialogue with national institutions and to raise awareness of the cross-cutting nature of the climate problematic.

IV   IN CONCLUSION

As any litigation, the climate cases discussed here tell stories: 'The energy industry is responsible for emissions causing climate change, this affects the enjoyment of ESC rights and the industry should pay damages'; 'The Government has not implemented proper regulations and policies and is failing

117   See ibid.
118   See Rodriguez-Garavito ibid.
119   See for instance Gauri & Brinks (note 116 above); Rodriguez-Garavito ibid.
120   See note 116 above.
121   Burns & Osofsky (note 34 above).

its duties to protect the commons it holds in trust, such as the atmosphere'. Legal cases are framed in adversarial and more or less simplistic terms, where only certain parts of the overall story are relevant in the courtroom. Legal narratives use purified roles of 'victim' (petitioner) and 'guilty party' (defendant) seeking to adjudicate responsibility and find corrective solutions. There is limited space for complexity or dual roles of actors as simultaneously causing and suffering harms. Despite these intrinsic limitations, the story-telling quality of court cases makes climate change more tangible and immediate, which significantly transforms the tone of the climate debate.[122]

When people who have actually been affected or are at imminent risk present their stories in a courtroom, the usual abstraction and ambiguity of climate change debates evaporates and makes room for discussing concrete reparations, prevention measures, and ways forward. In this chapter we have seen examples of litigation that succeeded in drawing attention to powerful climate change narratives that –regardless of success in court– captured public attention and imagination. The cases of *Kivalina,* the *Inuit Circumpolar Conference* and *the Maldives* are paradigmatic cases around which intense climate debates have grown, drawing attention to harms that otherwise could remain out of public discussion, and combining 'green' and 'red' concerns by putting names and faces to concrete injuries faced by individuals and groups who otherwise would remain abstract. Yet, allocating climate responsibility through courts may bring too little relief too late.[123]

While this chapter has aimed to show the potential that litigation harbours for addressing climate change related problems, it is clearly no panacea. Radical critics of rights based strategies, including climate change litigation, see them as demobilizing and leading up a blind alley, where contestation is circumscribed by premises defined by the dominant social and economic powers. Patrick Bond and Khadija Sharife's chapters in this volume exemplify this perspective, arguing that the terms of the struggle must be redefined to meaningfully address the problems of climate change and climate justice, and offering the concept of commons as a more promising path to radical transformation.[124] Such criticisms should be taken seriously. Nonetheless, the discussions in this chapter indicate that litigation strategies are not necessarily demobilizing, but may complement and mobilize other forms of activism, and that litigation does harbour a potential for contributing towards transformation and structural change. On the other hand, our analysis also points to limitations of climate change litigation strategies.

Climate change related problems present significant challenges –not only for political governance, but also to legal systems–. The science involves

---

122 Hunter (note 1 above).

123 International Council on Human Rights Policy (ICHRP) *Climate Change and Human Rights: A Rough Guide* (2007). See also HM Osofsky 'The Continuing Importance of Climate Change Litigation' (2010) 1 *Climate Law* 15.

124 P Bond 'Water rights, commons and advocacy narratives' (2013) in this volume; K Sharife & P Bond 'Payment for ecosystem services versus ecological reparations: The "green economy", litigation and a redistributive eco-debt grant'(2013) in this volume.

complexities of global political ecology that are of a scale new to the courts.[125] Cimate lawfare has sought, among other to improve assessment of climate impacts and use climate science for socio-legal debate; improve consideration of climate impacts in project planning and public financing decisions; compel public agencies to regulate CO_2 and other GHGs as pollutants emission; set reduction targets; put pressure on fossil-fuel corporations to revise their technologies and investments; secure compensation for affected populations; and review risks posed to world heritage sites arising from climate change.[126] Some of these strategies have had a certain amount of success for plaintiffs, but the courts have also rejected many.

While litigation based on administrative law in some cases has succeeded in changing climate related public policies, international legal institutions have not accepted red-green international law arguments –nor have tort-based cases been successful–. The failure of climate tort lawfare in court seems to stem from the frame in which civil law is designed; with strict causation, jurisdiction and addressability of harms requisites unsuitable for global, long term, and cross-cutting injuries. Hence, climate litigation seems to show an interaction between regulatory scale, science and law.

A criticism of climate litigation has been that the wide variety of strategies and legal regimes used blurs the issue of climate change, and makes the legal environment unclear for all parties: lawyers, governments, emitters and affected communities.[127] However, this discussion has shown that different legal strategies run up against different challenges, and have different potential. Hence a pluralistic strategy is not only what is likely to continue, but may also have advantages.

We have seen how the unique aspects of climate change 'have forced climate advocates to innovate and to develop creative new strategies internationally and domestically'.[128] Looking at emerging and potentially new red-green lawfare strategies, we have drawn attention to how courts (particularly in countries of the global south) in cases regarding social, economic and cultural rights, have developed alternatives to the political question doctrine that permit them to act without designing public policy. In cases on housing, public healthcare, and environmental protection among others, 'judicial experimentalism'[129] has resulted in dialogic or structural judgments where the court, finding violations of social, economic, cultural or environmental rights, have ordered the executive and/or legislative branches to design suitable responses. Often this is combined with follow-up mechanisms to evaluate compliance with the rulings.[130] The scarcity of substantial climate change litigation in the global south makes it difficult to pose this as an alternative approach, but it serves to illustrate that the political question approach is neither universal

---

125 Hunter (note 1 above).
126 Ibid.
127 Ibid.
128 Ibid.
129 See for instance P Bergallo 'Justice and Experimentalism: Judicial Remedies in Public Law Litigation in Argentina' (2005) *Society for Entrepreneurship in Latin America 2005.*
130 Rodríguez-Garavito (note 116 above).

nor inevitable.[131] Hence cases filed elsewhere could possibly have different outcomes.

The cases discussed in this chapter to varying degrees combined 'red' and 'green' concerns. 'Greener' lawfare has typically played out in cases focusing on the precautionary principle, and environmentally sustainable development. At the core of the narratives presented in these cases is the increasing scientific understanding of the impacts of human industrial activity on the planet's climate – and by implication on people's lives—. These cases do not ignore socio-economic rights, but the core 'red' issues of (transnational and intergenerational as well as domestic) climate justice, is not a strong narrative.

To our knowledge, no court case has used the international climate change legal framework to challenge states to give special protection to vulnerable communities against harmful climate events. Nor have arguments been raised to challenge unequal access to energy resources. In many of the cases where the 'red' narrative has been prominent, courts have reacted negatively. This includes cases brought concerning indigenous communities who are particularly vulnerable or whose rights are violated due to climate change.

Nonetheless climate litigation has had an important role in making affected communities more visible in domestic and international forums. The effectiveness of red-green climate lawfare is thus difficult to determine. Notions of effectiveness vary according to the goal set by those taking the case to court.[132] We have seen that even failed cases have had indirect political effects – and these may in some cases be what the litigants were primarily aiming for—. One indirect effect of climate litigation is that authorities and companies increasingly must be aware of these problems and anticipate the impacts of their action, as their liability is at risk.[133] Another indirect effect of these cases has been to place and keep the issue on the agenda. Besides functioning as adjudicators of disputes and aiding the implementation social reforms, courts also serve as forum of protest and political debate 'where political and social movements communicate and generate support for their legal and political agenda.'[134] The indirect effects of litigation may thus constitute an important reason behind, and argument for climate change litigation.

131 The debate is not that smooth amongst USA jurists either. For contrasting perspectives on the political question doctrine in the USA, see for instance Chemerinsky (note 53 above); Thorpe (note 53 above); Tribe et al (note 55 above).

132 Lin (note 24 above).

133 Following Osofsky, in this context and in particular the links between climate issues and socio-economic vulnerability are becoming clearer, new actors are becoming involved. International financial institutions and insurance systems may contribute to environmental risk management. In the context of lawsuits, insurers are made to work on conflict resolution, hence they need to develop expertise and provide new funding for the valuation of climate change effects and liabilities. HM Osofsky 'Local Approaches to Transnational Corporate Responsibility: Mapping the Role of Subnational Climate Change Litigation' (2007) 20 *Global Business & Development Law Journal* 143.

134 Lin (note 24 above) 49, building on J Lobel 'Courts as Forums for Protest ' (2004-5) 52 *UCLA L Rev* 477.

AUTHOR BIOGRAPHIES

Anna Alcaro currently works in the Office of Global Strategy and Programs (OGSP) at Duke University (USA), where she provides administrative support for the Africa Initiative. She is a Pennsylvania native and, prior to working at Duke, Anna was a Fulbright Fellow in Johannesburg, South Africa. At that time (2011-12), she worked at the Socio-Economic Rights Institute of South Africa (SERI), where her research focused on themes of poverty and socio-economic rights and their interplay with the environment, specifically related to mining and industrial pollution. Prior to working at SERI, Anna received her BA in Anthropology from The Pennsylvania State University. Her most recent publication – together with Jackie Dugard and Jennifer MacLeod – is 'A Rights-Based Examination of Residents' Engagement with Acute Environmental Harm across Four Sites on South Africa's Witwatersrand Basin' (2012) 79(4) *Social Research* 931-956.

Patrick Bond is a political economist at the University of KwaZulu-Natal in Durban, South Africa, where he directs the Centre for Civil Society (http://ccs.ukzn.ac.za). His work presently covers economic crisis from global to local scales; environment especially energy, water and climate change; social mobilization; public policy; and geopolitics. His publications are grounded in South Africa, Zimbabwe, the African continent and global-scale processes. Amongst his books are *Politics of Climate Justice* (2012); *Durban's Climate Gamble* (2011); *Zuma's Own Goal* (2010) and *Climate Change, Carbon Trading and Civil Society* (2009).

Brandon Barclay Derman is a doctoral candidate in Human Geography and a graduate fellow in the Comparative Law and Society Studies Center at the University of Washington (UW) (USA). His dissertation research examines advocacy and activism for social justice in relation to the uneven geographies of climate change. While at UW, Brandon has also served as a teaching and research fellow for the Three Degrees Climate Justice Project in the School of Law. His recent publications include S Herbert, B Derman & T Grobelski 'The Regulation of Environmental Space' (forthcoming 2013) 9 *Annual Review of Law and Social Science*. He holds a BA from the University of Michigan (French, Contexts of Urban Design), and an MA from Hunter College, City University of New York (Geography, Outstanding Master's Thesis).

Jackie Dugard is a visiting senior fellow at the School of Law, University of the Witwatersrand (South Africa). She is also a senior researcher at the Socio-Economic Rights Institute of South Africa (SERI), which she co-founded in January 2010 and headed until December 2012. She has a background in law and social sciences and tries to combine academic and activist tendencies. Her research, publication and advocacy interests include the role of law and the courts in social change, as well as socio-economic rights. Among her recent publications are the forthcoming *Socio-Economic Rights in South Africa: Symbols or Substance?* (2013), which she co-edited with Malcolm

Langford, Ben Cousins and Tshepo Madlingozi and the co-authored article, with Jennifer MacLeod and Anna Alcaro, 'A Rights-Based Examination of Residents' Engagement with Acute Environmental Harm across Four Sites on South Africa's Witwatersrand Basin' (2012) 79(4) *Social Research* 931-956.

Kjersti Fløttum is Professor of French linguistics at the Department of Foreign Languages, University of Bergen (Norway), where she teaches text theory and discourse analysis. She headed the Bergen Summer Research School (BSRS) on Global Development Challenges 2008-2011. Her general research fields are related to text and genre theory with a special focus on linguistic polyphony (multivoicedness) and dialogism in scientific, political and climate change discourse. She is head of the multidisciplinary project LINGCLIM: Linguistic representations of climate change discourse and their individual and collective interpretations (2013-2015), investigating the role of language in the climate change debate on national and international levels. Fløttum is co-author of the book K Fløttum, T Dahl & T Kinn *Academic Voices* (2006); editor of K Fløttum *Speaking to Europe* (2013) and has published extensively in international journals including *Journal of Pragmatics*, *Journal of Language and Politics*, *Language & Communication* and *Climatic Change*.

Des Gasper works at the International Institute of Social Studies in The Hague (Netherlands), a graduate school within Erasmus University Rotterdam. He studied economics, international development and policy analysis at the universities of Cambridge and East Anglia, and worked through the 1980s in Africa as a government planner and university lecturer. His research in recent years has involved theorising around well-being, human security and development ethics, with applications especially in areas of migration and climate change. Recent publications include *Development Ethics* (co-editor, 2010); *Trans-National Migration and Human Security* (co-editor, 2011); 'Climate Change – The Need For A Human Rights Agenda Within A Framework Of Shared Human Security' (2012) 79(4) *Social Research*; *Migration, Gender and Social Justice* (co-editor, 2013) and 'Climate Change and the Language of Human Security' (2013) 16(1) *Ethics, Policy and Environment*.

Øyvind Gjerstad has a Phd in French linguistics and discourse analysis from the University of Bergen (Norway). He has worked mainly on the polyphony, or 'multivoicedness', of political discourse. Recently he has taken interest in the narrative properties of climate change discourse, which often comprises a plot represented by the problem of climate change and the different social actors' (re)actions during the unfolding story. He is currently a postdoctoral fellow in the project LINGCLIM at the University of Bergen.

Siri Gloppen is Professor of Comparative Politics at the University of Bergen (Norway), Senior Researcher at the Chr Michelsen Institute, and a research coordinator at the PluriCourts Center of Excellence, University of Oslo. Her research focuses on the role of law and legal institutions in social and political transformation, spanning climate change lawfare, legal mobilisation and enforcement of social rights, constitution-making, judicial politics, and transitional justice. She has directed a number of comparative research projects in these areas, resulting in a several books and articles including *Law and Power in Latin America and Africa* (2010) and *Litigating Health Rights: can courts bring more justice to health?* (2011).

Jill Johannessen is a post-doctoral student at the Department of Information and Media Science at the University of Bergen, Norway. Her post-doc research is part of a larger project 'Climate Crossroads: Towards Precautionary Practices: Politics, Media and Climate Change', financed by the Norwegian Research Council. Johannessen's research examines how South African and Norwegian news media construct relations between climate, poverty, and responsibility between developed and developing nations, with the climate summit COP17 in Durban as the joint starting point. Previously, Johannessen worked as a Communication Leader at the Bjerknes Centre for Climate Research (BCCR) for five years. Johannessen has a Doctorate in Sociology, with specialization in Mass Communication, from the Norwegian University of Science and Technology. She has several publications on the intersection between gender, media, women's movement and development and fieldwork experience from Africa. She now specialises in cross-disciplinary research within media, climate and development.

Molefi Mafereka Ndlovu has a degree in Community and Development Studies from the University of KwaZulu-Natal (UKZN) (South Africa). He joined the UKZN Centre for Civil Society (CCS) as a researcher in the Energy Project, where he was concerned with water privatisation in Johannesburg and linking this to the Mega-Dams development in the Lesotho Highlands Water Project. He has presented his work at conferences in Lesotho and Korea and was published in the book *Development Dilemmas In Post-Apartheid South Africa* (2010) edited by Bill Freund and Harald Witt. His recent research has focused on the issue of voice and self-representation by affected groups in society. Since 2011, Molefi has been working in collaboration with other artists and community media activists on building an online digital media archive. 'Qwasha!: Climate Change' is one of the thematic areas of the archive, in which rigorous scholarship sits alongside creative inputs and records from the margins of recognised society, combining different strategies for knowledge production in conversation with each other, online.

Ana Victoria Portocarrero is a Nicaraguan feminist economist, with a Master's degree in Development Studies from the International Institute of Social Studies in The Hague. She is currently working on the intersections between climate change policies, food sovereignty and trade, focusing on the Central American Region. She is particularly interested in the philosophical and political economy dimensions of the discourses and practices of climate change policy, and the ways in which these logics justify and perpetrate the social and economic exclusion of minoritised groups. Her most recent publication on climate change, co-authored with Des Gasper and Asunción Lera St. Clair, is 'The framing of climate change and development: A comparative analysis of the Human Development Report 2007/8 and the World Development Report' (2012) 23 *Global Environmental Change* 28-39.

Khadija Sharife is a post-graduate law student, coordinator for the Environment Justice Trade and Liabilities (EJOLT) project at the Center for Civil Society (CCS), University of Kwa-Zulu Natal, based in South Africa, and commissioning editor at the Forum for African Investigative Reporters (FAIR), Africa's primary investigative journalism organization. She is an Africa project fellow at the US-based World Policy Institute, assistant Africa editor of *Capitalism, Nature, Socialism,* and lead author of *Tax Us if You Can: Africa* (2011) in addition to other books. An investigative journalist and correspondent for *African Business Magazine, Africa Report* and others, her work has appeared in *African Banker, New African, Al Jazeera, Forbes, BBC, Le Monde Diplomatique, The Economist, Harvard International Review, London Review of Books,* as well as academic journals including *ROAPE.* She is a board member of the Center for Natural Resource Governance (CNRG), and former board member of FAIR.

Asunción Lera St. Clair, philosopher and sociologist, is Research Director at the International Centre for Climate and Environmental Research-Oslo (CICERO) (Norway), former Professor of Sociology at the University of Bergen, and Associated Senior Researcher at the Chr Michelsen Institute in Bergen. St. Clair is Lead Author of the Fifth Assessment Report of the Intergovernmental Panel on Climate Change (IPCC) for the Working Group II *Report on Impacts, Adaptation and Vulnerability;* member of the Joint Programming Initiative for Climate (JPI Climate). Her research interests are focused on the relations between transformative change, critical poverty studies, climate change, development ethics, human rights and global justice, with a particular focus on epistemology and processes of knowledge production; and the role of donor and multilateral organizations.

Petra Tschakert is an Associate Professor of Geography and the Earth and Environmental Systems Institute (EESI) at the Pennsylvania State University in the USA and a Senior Research Fellow with the Center for International Climate and Environmental Research Oslo (CICERO) in Norway. She is Coordinating Lead Author on Chapter 13 'Livelihoods and Poverty' of the Fifth Assessment Report of the Intergovernmental Panel on Climate Change (IPCC), Working Group II on Impacts, Vulnerability, and Adaptation. Tschakert is an internationally recognized scholar working at the intersection of political ecology, climate change adaptation, social-ecological resilience, environmental justice, livelihood security, and participatory action research and learning within a development context. She directs several research initiatives that explore and facilitate co-generative inquiry, adaptive capacity, and livelihood resilience among resource-poor men and women land users in Africa and the Himalayas. Her work on adaptation is published in *Global Environmental Change; Ecology & Society; Ethics & Social Welfare; Emotion, Space & Society* and *Climatic Change.* In collaboration with Nancy Tuana and Carolyn Sachs, she is currently working on a book on gender and climate change, to be published in 2014.

Nancy Tuana is the founding director of Penn State's Rock Ethics Institute and Professor of Philosophy and Women's Studies at Penn State University (USA). She is a philosopher of science who specialises in issues of ethics and science, as well as feminist science studies. Nancy is the principle investigator on two interdisciplinary research projects funded by the National Science Foundation that include a network for Sustainable Climate Risk Management, linking a transdisciplinary team of scholars at 19 universities and 5 research institutions across 6 nations, to answer the question 'What are sustainable, scientifically sound, technologically feasible, economically efficient, and ethically defensible climate risk management strategies?' Nancy also leads a research initiative on gender and climate change supported by the Worldwide Universities Network (WUN). Her work has been published in publications such as *Climatic Change*; *Synthese*; *Ethics, Policy and Environment* and *Hypatia*. In collaboration with Petra Tschakert and Carolyn Sachs, she is currently working on a book on gender and climate change, to be published in 2014.

Catalina Vallejo is a lawyer from the Universidad Autónoma Latinoamericana (Colombia). She has specialised in administrative law and holds a Master's degree in peace studies, which she obtained from Innsbruck University in Austria. She has worked for the Colombian public sector in projects related to land management, urban planning and human rights. Vallejo currently collaborates with the Chr. Michelsen Institute (Norway) on various research projects with regional focus on Latin America, including studies on transitional justice, civilian-military relationships, and climate change lawfare. She is the author of *Plurality of peaces in legal action: analysing constitutional objections to military service in Colombia* (2012).

TABLE OF CASES

INDEX

www.ingramcontent.com/pod-product-compliance
Lightning Source LLC
Chambersburg PA
CBHW080644270326
41928CB00017B/3184